Radiation Dose Management of Pregnant Patients, Pregnant Staff and Paediatric Patients

Diagnostic and interventional radiology

IPEM–IOP Series in Physics and Engineering in Medicine and Biology

About the Series

Series in Physics and Engineering in Medicine and Biology will allow IPEM to enhance its mission to 'advance physics and engineering applied to medicine and biology for the public good.'

Focusing on key areas including, but not limited to:

- clinical engineering
- diagnostic radiology
- informatics and computing
- magnetic resonance imaging
- nuclear medicine
- physiological measurement
- radiation protection
- radiotherapy
- rehabilitation engineering
- ultrasound and non-ionising radiation.

A number of IPEM–IOP titles are published as part of the EUTEMPE Network Series for Medical Physics Experts.

Radiation Dose Management of Pregnant Patients, Pregnant Staff and Paediatric Patients

Diagnostic and interventional radiology

Edited by
John Damilakis
University of Crete, Faculty of Medicine, Department of Medical Physics, Iraklion, Crete, Greece

IOP Publishing, Bristol, UK

ISBN 978-0-7503-1317-9 (ebook)
ISBN 978-0-7503-1318-6 (print)
ISBN 978-0-7503-1785-6 (myPrint)
ISBN 978-0-7503-1319-3 (mobi)

DOI 10.1088/978-0-7503-1317-9

Version: 20191201

IOP ebooks

British Library Cataloguing-in-Publication Data: A catalogue record for this book is available from the British Library.

Published by IOP Publishing, wholly owned by The Institute of Physics, London

IOP Publishing, Temple Circus, Temple Way, Bristol, BS1 6HG, UK

US Office: IOP Publishing, Inc., 190 North Independence Mall West, Suite 601, Philadelphia, PA 19106, USA

Contents

Preface

A 25-year-old woman, pregnant with a fetus at 15 weeks was admitted to the hospital to rule out appendicitis. The appendix was not visualized adequately with ultrasound. For this reason, clinicians requested an abdominal CT examination. Is the risk of irradiation of the conceptus less than that of not making the necessary diagnosis?

A 22-year-old female patient underwent a fluoroscopically-guided cardiac ablation procedure. The total fluoroscopy time was 42 min. One week after the procedure, the patient was diagnosed as pregnant, in her sixth week of gestation. Can we recommend abortion?

A 30-year-old interventional radiologist declares her pregnancy at the eighth week of gestation. She asks about embryo radiation doses from her professional activity, associated radiogenic risks and maximum workload for the next seven months to keep her baby safe. Can we provide well-informed advice?

A 15-year-old adolescent was admitted to the hospital with dyspepsia. Clinicians decided to request a barium meal examination. Can we estimate the effective dose to the patient?

This book provides information about the practical use of new information in radiation dose management of pregnant patients, pregnant staff and paediatric patients in diagnostic and interventional radiology. Topics range across medical dosimetry, biological effects of exposure to ionizing radiation during gestation and childhood, parameters that influence conceptus and paediatric dose, methods to calculate conceptus and paediatric dose and strategies to optimize examinations performed on pregnant and paediatric patients. The last chapter focuses on the management of pregnant patients and pregnant employees in diagnostic and diagnostic and interventional radiology. I hope, this book will be a useful resource for medical physicists, diagnostic radiologists, interventional radiologists and other physicians performing fluoroscopically guided procedures, radiographers, gynecologists, obstetricians and paediatricians.

Professor John Damilakis
November 2019

Editor biography

John Damilakis

John Damilakis is Professor and Chairman, Faculty of Medicine, University of Crete and Director of the Medical Physics Department, University Hospital of Iraklion, Crete, Greece. He is Vice President and President-elect (Vice President 2018–21, President 2021–24) of the International Organization for Medical Physics (IOMP), Immediate Past President of the European Federation of Organizations for Medical Physics (EFOMP) and Immediate Past President of the European Alliance for Medical Radiation Protection Research (EURAMED).

John Damilakis has published more than 250 research articles in peer-reviewed journals and 206 research articles in journals cited in PubMed (October 2019). Many of these publications are in leading journals such as *Medical Physics*, *Physics in Medicine and Biology*, *Radiology*, *Investigative Radiology* and *European Radiology*. His main research interests include the development of methods for accurate assessment of radiation dose from medical examinations and associated risks with a special interest in embryo/fetal dosimetry, paediatric dosimetry, computed tomography and interventional radiology dosimetry, investigation of radiation dose reduction strategies in diagnostic and interventional radiology and optimization of procedures with emphasis on radiation dose and image quality.

The goals of his research work have mostly been derived from his everyday practice as a clinical medical physicist. He is best known for his pioneering research work on the estimation of conceptus dose and radiogenic risk associated with diagnostic and therapeutic radiological procedures. His papers have become the classical reference on the conceptus dosimetry for interventional and diagnostic imaging. A software package (COnceptus Dose Estimation, CODE) was developed in 2014 by his research team and became a very popular tool worldwide. CODE is a free web-based (http://embryodose.med.uoc.gr/code/) software tool developed for the estimation of conceptus radiation dose and radiogenic risks in case of: (a) pregnant patients subjected to radiological examinations and (b) pregnant employees exposed during fluoroscopically guided interventional procedures.

Professor Damilakis has been invited to give numerous talks and keynotes on medical dosimetry and medical radiation protection topics at international scientific events around the world. Of special note is his role as project leader in several projects in the field of medical radiation protection and medical dosimetry. He has received many awards and distinctions in recognition of his academic accomplishments.

List of contributors

J Damilakis
Conceptus dose and radiation-induced risk associated with diagnostic and interventional X-ray examinations (4.1); Methods to calculate conceptus dose from diagnostic and interventional procedures (5.1); The management of pregnant patients (8.1); The management of pregnant employees (8.2).

A Papadakis
Thermoluminescence dosimeters (TLDs) and optically stimulated luminescence dosimeters (OSLDs) (1.2); Optically stimulated luminescence dosimeters (OSLDs) (1.3); Radiography and fluoroscopy parameters that influence conceptus and paediatric dose (3.1); Optimization of radiographic and fluoroscopic examinations performed on paediatric patients (7.1.5, 7.1.6 and 7.1.8).

K Perisinakis
Biological effects to conceptus from ionizing radiation (2.1); CT parameters that influence conceptus and paediatric dose (3.2); CT during pregnancy: methods for dose optimization (6.2).

J Stratakis
Physical phantoms simulating pregnancy and children (1.1); Computational phantoms simulating pregnancy and children (1.4); Monte Carlo simulation (1.5); Radiography/fluoroscopy during pregnancy: methods for dose optimization (6.1); Optimization of radiographic and fluoroscopic examinations performed on paediatric patients (7.1.1–7.1.4 and 7.1.7).

V Tsapaki
Biological effects to children from ionizing radiation (2.2); Paediatric dose and radiation-induced risk associated with diagnostic and interventional X-ray examinations (4.2); Methods to calculate paediatric dose from diagnostic and interventional procedures (5.2); Methods of dose optimization in CT (7.2); appendix.

Chapter 1

Dosimetry

John Stratakis and Antonios Papadakis

Soon after the discovery of X-rays, the necessity for radiation dosimetry rapidly became apparent, especially after the harmful effects of radiation were realized. Physical or mathematical phantoms are entities that represent the human anatomy and currently are used for dosimetry purposes. Most of the physical phantoms have explicit properties and have been manufactured to simulate tissues for specific measurements in diagnostic or treatment energy ranges of ionizing radiation. In addition, mathematical phantoms have become extremely popular and are used in all parts of radiation physics applications. Direct dosimetric methods using either a thermoluminescence (TL) technique or optically stimulated luminescence (OSL) technique or any other method using passive solid-state detectors, have also been used for the estimation of absorbed doses for both patients and personnel in radiotherapy, diagnostic radiology and nuclear medicine applications.

1.1 Physical phantoms simulating pregnancy and children

1.1.1 Introduction to physical phantoms

Initially, phantoms were constructed from water, wax or other uncomplicated materials. While water is a good approximation of soft human tissue, wax and other materials demonstrated difficulties in simulating tissue equivalency, since the properties of phantom materials may vary with the energy of incident radiation. Furthermore, simulated tissues may have diverse properties, both geometrical and radiological and the goal to represent these physical and radiological properties is an arduous challenge. New tissue substitutes such as resins, polystyrene and polyurethanes have allowed the construction of realistic anthropomorphic phantoms and improved their reproducibility and their tissue simulation capabilities in a wider range of energies. Varieties of phantoms exist, whilst a simple geometric exterior can

doi:10.1088/978-0-7503-1317-9ch1

contain a variety of planar or three-dimensional inhomogeneities as detailed internal structures, simulating muscle, bone and lung tissue.

1.1.2 Phantom materials

Selecting the appropriate materials for phantoms simulating the geometry and the composition of the human body, is critical and directly related to the design, purpose and cost of any type of phantom. The first phantoms were relatively modest and unsophisticated, consisting of slabs of simple materials such as wax or water. However, the dominant goal of anatomical physical phantoms, to reproduce the attenuation and scattering properties of the human body as precisely as possible, made many of early materials quickly fall out of use since their properties could not be used as a measure of the tissue equivalence inside the phantoms.

Physical density and the mass attenuation coefficient are physical properties that can be used as basic parameters of a material's tissue equivalence. Nonetheless, problems can arise since mass attenuation coefficient is energy dependent and physical density cannot effectively describe a material's radiological properties; how the material will behave to incident radiation. More detailed parameters that provide insight into how phantom materials behave in a radiation field are photon interaction parameters such as mass energy absorption coefficient, effective atomic number (Z_{eff}) and effective electron density. Therefore, an ideal material is one that will more accurately represent these interaction parameters of the specific human tissue material being simulated (Hill *et al* 2005). Consequently, the selection of tissue simulants is based on data providing the elemental composition of each tissue being simulated (ICRU 1992, 2016).

Molecular compounds mixed with epoxy resins are used for phantom construction in order to achieve the desired density and mass attenuation coefficient. The effective atomic number and density of the compound are adjusted to the desired values, by adding constituents of – chemically inert – phenolic hollow spheres. The composition of these epoxy compounds is often proprietary information of each manufacturer and the end-user products are frequently stated to follow ICRU recommendations in terms of radiation and non-radiation related properties (attenuation, scatter, thermal and mechanical properties). Usually, a single soft-tissue material is used to simulate all human soft tissues (1.04 g cm^{-3}) and a similar compound with almost identical elemental composition but with lower density (0.3–0.35 g cm^{-3}) is assigned to the lung tissue. Soft tissue compounds are used with foaming agents to achieve lower densities of lungs. Bone density (1.4–1.6 g cm^{-3}), on the other hand, is accomplished using a variety of methods. Some phantoms contain real bone structures or molded bone material extracted from cadavers. While the first method can provide precise anatomy, a human cadaver skeleton demonstrates reduced bone marrow and water content (Somerwil 1977). The molded bone equivalent approach solved the reduced bone density due to loss of marrow and problems arising due to continuity of bone mimicking structures immersed into the soft-tissue epoxy-resin matrix. Reduced density regions, such as adipose tissue, which possess a lower density than soft tissue, are usually not simulated. However, specific phantoms can utilize mixtures of adipose and/or glandular equivalent materials to simulate breast tissue and localized external fat.

Apart from empirically testing material constituents for matching attenuation and density characteristics of tissue-mimicking materials, it is also frequently necessary to validate the appropriateness of substitutes with available tabulated data for cross-sectional, attenuation and absorption coefficients either published, or available online through repositories or computer programs (Hubbell and Seltzer 2004, Berger *et al* 2011, Boone and Chavez 1996, Nowotny 1998, ICRP 1975, ICRU 1992, Taylor *et al* 2012, Hubbell 1982).

1.1.3 Anthropomorphic phantoms

The term 'anthropomorphic' is defined as the attribution of human form and properties to a non-human entity or object. Anatomically anthropomorphic physical phantoms are designed to simulate human geometries and closely resemble the shape and size of the human male, female and child. Even if commonly stated that their mass and height conform to 'Reference Man' (ICRP 1975, ICRP Publication 89 2002), anatomical phantoms can vary and do vary, just like individual humans. This is due to several reasons. Phantoms that include skeleton parts from cadavers are reliant on the original skeleton dimensions, in conjunction with those that utilize synthetic materials. The latter demonstrate more constant dimensions between phantoms. In addition, phantoms may reflect anatomic and geometric properties of the actual population they are intended to simulate. As a result, locations of an individual organ cannot be accurately defined. Therefore, even if the phantom can accommodate many dosimeters, which suggests a detailed dose map throughout its structure, the individual organ dose calculations may not always reflect actual organ doses due to uncertainties in specific locations. Several series of anthropomorphic phantoms are commercially available, while others have been constructed within institutions for specific research purposes.

1.1.4 Commercially available anthropomorphic phantoms

A complete range of anthropomorphic phantoms are commercially available from Computerized Imaging Reference Systems (CIRS, Norfolk, VA). The CIRS phantom range consists of a full line of anthropomorphic, cross-sectional dosimetry phantoms designed to investigate organ dose, whole body effective dose as well as verification of delivery of therapeutic radiation doses. Six models are available: newborn, 1, 5 and 10 years old paediatric phantoms as well as adult male and female phantoms. Characteristics of the tissue-equivalent materials used in CIRS phantoms are tabulated in table 1.1. Each phantom is sectional in design with traditional 25 mm thick sections and manufactured including tissue-equivalent materials representing bone, lung and soft tissue. CIRS ATOM® phantoms are formulated with synthetic epoxy-based bone equivalent materials based on the appropriate bone composition typical of each age (table 1.1). The skeletal anatomy of the ATOM® phantom is based on a homogeneous bone tissue composition that averages known cortical to trabecular bone ratios and age-based mineral densities. For most organ

Table 1.1. Physical and radiological parameters of commercially available physical phantoms.

Phantom	Height (cm)	Weight (kg)	Soft tissue d (g cm^{-3})	Soft tissue Z_{eff}	Bone tissue (g cm^{-3})	Bone tissue Z_{eff}	Lung tissue (g cm^{-3})	Lung tissue Z_{eff}
Reference Man	176	73	0.987	6.86	1.486	8.75	0.296	7.14
RANDO	175	73.5	0.998	7.6	Natural skeleton		0.352	7.11
Phantom	**Height (cm)**	**Weight (kg)**	**Soft tissue d (g cm^{-3})**	**Soft tissue Z_{eff}**	**Bone tissue (g cm^{-3})**	**Bone tissue ED × 10^{23}**	**Lung tissue (g cm^{-3})**	**Lung tissue Z_{eff}**
ATOM–0 year	51	3.5	1.055	7.15	1.41	4.498	0.210	7.38
ATOM–1 year	75	10	1.055	7.15	1.45	4.606	0.210	7.38
ATOM–5 year	110	19	1.055	7.15	1.52	4.801	0.210	7.38
ATOM–10 years	140	32	1.055	7.15	1.56	4.878	0.210	7.38
ATOM–adult female	173	73	1.055	7.15	1.60	5.030	0.210	7.38
ATOM–adult male	160	55	1.055	7.15	1.60	5.030	0.210	7.38

dosimetry applications, bone tissue can be reasonably simulated using homogenized bone properties. In addition, and because scattering radiation can give a significant dose contribution to the surrounding tissue, the ATOM® phantom newborn and 1 year old are provided with arms and legs as a standard configuration. However, extremities can be ordered separately for most standard ATOM® models.

CIRS phantoms utilize epoxy-based materials that, according to the manufacturer, can be machined to achieve optimal flatness, thus the air interface between sections is minimized.

Each phantom comes as standard with a unique holding apparatus that consists of a top and bottom compression plate that can be adjusted. The 25 mm plates are joined with Teflon wires that are radiographically opaque. There are no solid rods running through the phantom body to interfere with the imaging characteristics. Instead, the top of each section has a small tab that fits into the corresponding indentation on the underside of the preceding section. The series of ATOM® phantoms implements organ hole locations optimized for precise calculations using the minimum number of detectors necessary. The selection of hole positions is supported by detailed anatomical information about the average position of these 22 radiosensitive internal organs. A set of maps outlining the most frequently observed organ locations and also the optimized detector hole distribution within each organ accompanies each phantom with '-D' configuration with extra charge. This map book shows the hypothetical outline of the internal organs appropriate for each section. The CIRS family have been recently utilized in conjunction with thermoluminescence dosimeters (TLDs) crystals or metal-oxide-semiconductor field-effect transistor (MOSFET) dosimeters to assess the dose to paediatric patients undergoing multiple detector computed tomography (MDCT) CT angiography and digital fluoroscopy examinations (Perisinakis *et al* 2006, Fricke *et al* 2003, Frush and Yoshizumi 2006, Mukundan *et al* 2007, Blinov *et al* 2005).

Since no phantom simulating a pregnant patient exists in the ATOM® family, the female phantom has been modified by researchers (Hurwitz *et al* 2006) to simulate the expected location of a fetus at the first trimester of gestation to assess the radiation burden to the conceptus from clinical MDCT body scans. CIRS, apart from leg, arm and breast attachments for male and female phantoms, offers optional fat 4 cm thick layers that can be applied to phantoms. These layers, which can be used in sets or separately, mimic human fat tissue within 2% by liner attenuation for kV MV^{-1} energies.

The RANDO® phantom closely simulates the original Alderson RANDO® phantom (Alderson *et al* 1962). RANDO® phantoms consist of 2.5 cm thick sections, and hole grids are drilled into the soft and lung tissue-equivalent areas for each slice to accommodate TLDs. Holes are never drilled in places where bones are located. The RANDO® phantoms incorporate air, human cadaver skeleton or simulated bone material, a soft tissue-equivalent material and a lung tissue-equivalent material. The RANDO® soft tissue-equivalent material is a urethane formulation and is designed to replicate the effective atomic number, electron density and physical density of muscle tissue with randomly distributed fat. The RANDO® lung tissue-equivalent material was designed similarly, replicating the effective atomic number, electron density and physical density of lung tissue in a median respiratory state (0.35 g cm^{-3}). The RANDO® phantom has been proven extremely useful in a handful of studies, to quantify organ and effective patient radiation doses. Theocharopoulos *et al* (2002) used the RANDO® phantom in combination with TLDs to gauge the effective dose for anterior–posterior abdominal radiography, PA chest radiography, PA head radiography and AP heart fluoroscopy and compared it with effective doses derived with other dose estimations methods. Numerous studies have evaluated organ and effective doses from fluoroscopy and CT protocols (Fitousi *et al* 2006, Perisinakis *et al* 2004, 2005) utilizing the RANDO® phantom, while others have utilized it as a gold standard to validate mathematical phantoms for dosimetry. In addition, RANDO® phantoms were used many times as a patient substitute and scatter source in studies for occupational dosimetry (Theocharopoulos *et al* 2006, Perisinakis *et al* 2016, Osei *et al* 2001).

1.1.5 Research-based anthropomorphic phantoms

A series of anthropomorphic physical phantoms have been created for research at the University of Florida (UFL) as a complementary companion to specific computational models, permitting correlation of organ dosimetry between direct measurements and computer simulations. The UFL phantom family has been constructed using urethane-based soft tissue-equivalent materials modified specifically to match the radiological parameters of tissues in the energy range of diagnostic X-rays. While the first phantoms were not anatomical physical phantoms but were built in accordance to medical internal radiation dose (MIRD) stylized computational phantoms, the second generation of UFL phantoms provided much more realistic anatomical details, again representing adult and paediatric patients of newborn and 1 year age. The anatomical details of these phantoms were prepared from 1 mm axial CT data sets of actual deceased patients that had near as possible identical dimensions to the computational phantoms. The physical phantoms were constructed with a 5 mm slice thickness using epoxy-resin tissue-equivalent compositions (White 1978) modified accordingly to represent the age-dependent density of paediatric tissues, incorporating materials equivalent to air, lung, soft and bone tissue. Stylized values for newborn, child and adult tissues differ slightly, so these phantoms encompass various substances so that the radiological equivalence is adjusted to the age-specific reference values. More recently, anatomical phantoms representing 'Reference Male/Female' (Reference Man/Woman) have been constructed (Hintenlang *et al* 2008). An automated machining system allowed accurate fabrication of molds of 5 mm axial sections. Soft and lung tissue-equivalent materials utilize a proprietary urethane-based tissue-equivalent material that provides a pliable material permitting integration of immediate read-out dosimeters (Bower *et al* 1998, Wang *et al* 2004). Bone tissue consists of a rigid epoxy-resin-based tissue-equivalent material also used in similar phantoms. Since the UFL adult and paediatric phantoms are identically matched to a computational series of phantoms, they also provide the capability to benchmark and compare the results of physical measurements with computational simulations.

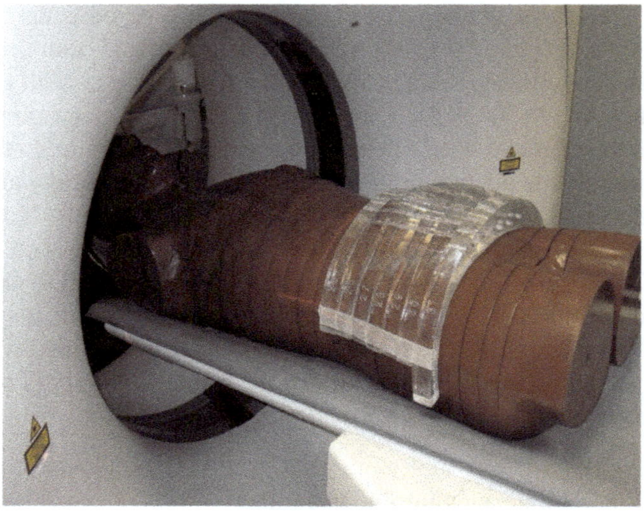

Figure 1.1. Phantom simulating pregnancy. Rings are fitted around the RANDO® phantom abdominal and pelvic sections to simulate pregnancy in the second trimester. Courtesy of the UOC.

The need for anthropomorphic physical phantoms simulating women in late pregnancy has urged many researchers to modify adult phantoms to determine radiation deposition from diagnostic procedures. The Medical Physics team of the University of Crete (UOC), School of Medicine, modified a RANDO® phantom consisting of 2.5 cm thick sections numbered 0 through 35 and additional rings constructed from Lucite ($C_5H_8O_2$, 1.19 g cm^{-3} in density) so that the phantom approximates, as nearly as possible, the attenuation characteristics of a pregnant woman (Damilakis *et al* 2000). To simulate pregnancy in the second trimester, seven 2.5 cm thick rings were fitted around sections 25 through 31 (figure 1.1); for a pregnancy in the third trimester, ten 2.5 cm thick rings were added to sections 22 through 31. Each section of the RANDO® phantom and each ring contained holes that permitted the placement of TLDs (figure 1.2). The anteroposterior dimensions and the width of each plane simulating pregnancy are shown in table 1.2.

1.2 Thermoluminescence dosimeters (TLDs) and optically stimulated luminescence dosimeters (OSLDs)

1.2.1 Introduction

Radiation dosimetric methods are used for the estimation of radiation dose absorbed using either the TL technique or OSL technique or any other method using passive solid-state detectors. These dosimetric methods have been used for the estimation of absorbed doses for both patients and personnel in radiotherapy, diagnostic radiology and nuclear medicine applications. The use of TL as a method for dosimetry of ionizing radiation has been well established and used for many decades in various fields (McKeever 1985, McKeever *et al* 1995, Kron 1994, McKinlay 1981, McKeever and Moscovitch 1995). The thrust of modern

Figure 1.2. An abdominal slice of the UOC phantom simulating pregnancy in the second trimester. TLD crystals indicate fetal positions. Courtesy of the UOC.

Table 1.2. Dimensions of the UOC modified RANDO® phantom sections simulating mid- and late pregnancy.

Second trimester phantom			Third trimester phantom		
Section no.	AP dimension (cm)	LAT dimension (cm)	Section no.	AP dimension (cm)	LAT dimension (cm)
25	24.0	30.0	22	23.5	310
26	25.0	30.0	23	27.0	31.0
27	26.5	30.5	24	30.0	30.0
28	27.5	31.5	25	32.5	30.0
29	27.5	33.0	26	33.0	30.5
30	27.5	34.0	27	33.5	31.0
31	26.5	34.0	28	32.5	32.0
			29	30.5	33.0
			30	27.5	33.5
			31	23.5	33.0

luminescence dosimetry evolvement is, however, more towards OSL (Bøtter-Jensen *et al* 2003). The purpose of this chapter is to provide an introduction to theory and use of TL and OSL for ionizing radiation dose measurement.

1.2.2 Luminescence – terminology

Some materials have the ability to absorb a certain type of energy, store a part of it for a period of time and dissipate it in the form of optical radiation. These materials are called luminescent materials and the process is called luminescence. Depending

Table 1.3. The different types of luminescence.

Luminescence effect	Means of excitation
Triboluminescence	Rubbing or grinding
Chemiluminescence	Chemical energy
Bioluminescence	Biochemical energy
Cathodoluminescence	Cathode rays
Photoluminescence	Optical photons
Radioluminescence	Ionizing radiation
Electroluminescence	Electric field
Sonoluminescence	Sound waves

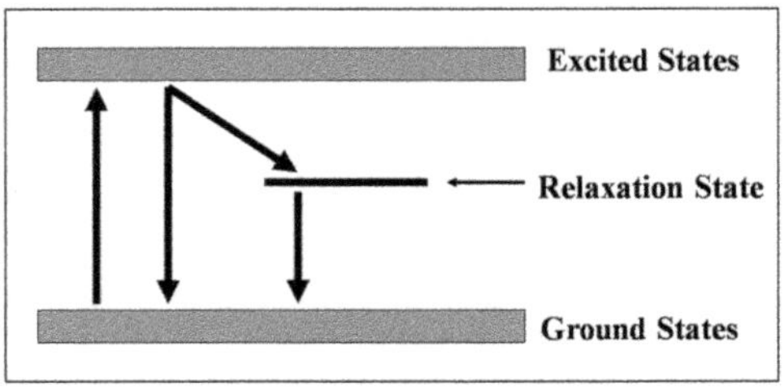

Figure 1.3. Fluorescence: Prompt return of an electron to the ground state either directly or via an intermediate relaxation state. De-excitation time <10 ns.

on the means by which the luminescent material receives the excitation energy, a prefix is added to the term luminescence (table 1.3). There is an exception to that rule causing confusion and misunderstanding about the luminescent phenomenon. This is the case when the excitation energy is heat.

A luminescent effect may be characterized as fluorescence or phosphorescence depending on the time elapsing after excitation until the emission of the optical radiation regardless of the means of excitation. Excitation occurs when a bound electron absorbs energy and is displaced to a higher energy state (excitation state). If the electron returns promptly either directly or via an allowed transition from an intermediate (relaxation) state to the initial energy state (ground state) with the excess in energy emitted as optical radiation, the process is called fluorescence (figure 1.3).

If there is an allowed energy state (metastable state) in the forbidden energy gap and transition from this metastable state directly to the ground state (valence band) is not allowed, energy should be provided to electrons in this metastable state to reach an excitation state (conduction band) and return to a ground state emitting the difference in energy as an optical photon. In such cases, with the return of the excited electrons to the valence band, consequently the emission of the optical photons is delayed by the metastable state and the process is called phosphorescence (figure 1.4).

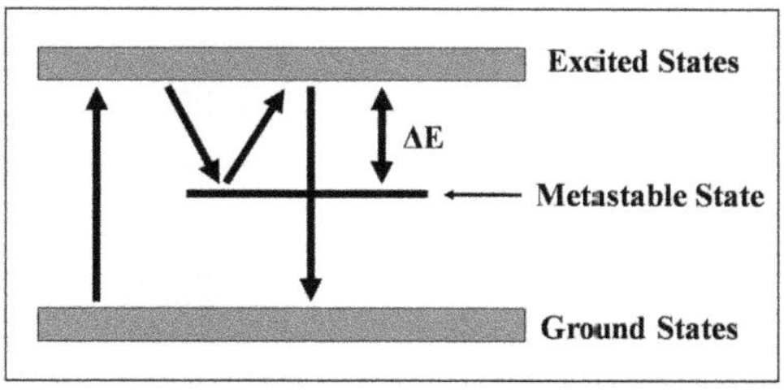

Figure 1.4. Phosphorescence. The return of an electron to the ground state is delayed by the metastable state since direct transition from this intermediate state to the ground state is not allowed. De-excitation time <100 ms.

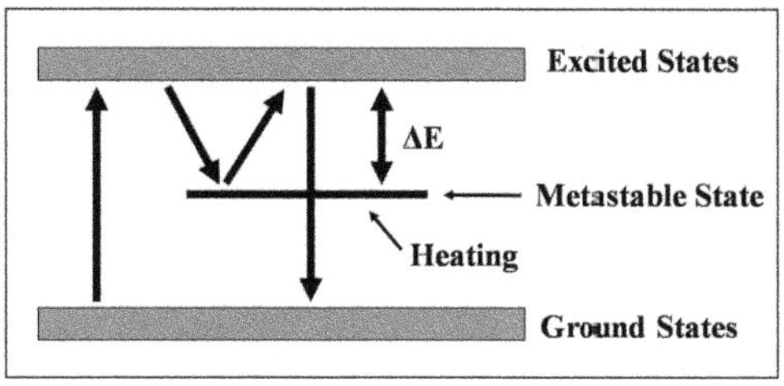

Figure 1.5. Thermoluminescence. The return of an electron trapped in the metastable state to the ground state is boosted by heating the material resulting in accelerated phosphorescence.

The probability *P* per unit time of an electron to escape the metastable state to an excited state is given by the Boltzmann's equation:

$$P = se^{(-\Delta E/kT)} \tag{1.1}$$

where *s* is a constant for a specific metastable state having dimensions of frequency, ΔE is the energy difference between the metastable state and the excited states, which are commonly called energy depth of the trap, *T* is the temperature of the material in K and *k* is the Boltzmann's constant. If we decrease the temperature of the luminescent material, the probability of electron escape increases. This accelerates the phosphorescence process, as electrons manage to escape more rapidly to excited states from progressively deeper metastable states (traps) and then return to ground states by emitting photons. The accelerated phosphorescence resulting after heating a luminescent material is called TL (figure 1.5). Thermoluminescence is a totally different process from incandescence. Light emission in incandescence results from the vigorous vibrations and collisions occurring in the heated material that lead to the excitation of all the atoms present. The emission spectrum in incandescence depends strongly on temperature, since the higher the temperature the more violent the vibrations and collisions are, similar to the black body emission spectrum at the same temperature. In contrast, the TL emission spectrum depends only on the type of atoms present in the material but not the temperature.

In all the above, if the excitation energy is provided by photons, the luminescence emission photons have lower energy than the excitation photons according to Stokes' law. There is another process of luminescence, mentioned here for completeness, the 'anti-Stokes' luminescence, which is the case when the emission photons have higher energy than those used for excitation. The principle of energy conservation is still valid since two or more excitation photons are absorbed for the excitation of an electron from the ground state to an excited state.

As mentioned above, fluorescence and phosphorescence have different timings. The question to be answered is how we characterize a luminescence process as fluorescence or phosphorescence. If the de-excitation of an electron raised in an excited state occurs within 10 ns after excitation the process is clearly fluorescence, whereas if the de-excitation persists for more 100 ms it is undoubtedly phosphorescence. In cases with intermediate de-excitation periods we decide among the two by examining the effect of rising temperature on the rapidity of the process, having in mind that heating accelerates the process of phosphorescence while has no effect on fluorescence. In the following text the term TL is used to describe the luminescent phenomenon produced exclusively as a result of the absorption of ionizing radiation.

1.2.3 Crystalline structure of thermoluminescent materials

Even though luminescence phenomena and especially TL has been under investigation since the very first steps of science, a theoretical model, fully explaining all the experimental results, has not been found. The mechanism of TL is very complex and although our understanding of the process has been highly improved over the last few decades, it has not been possible to produce a much better phosphor than LiF:Mg:Ti, which was the end product of a chance discovery in the late 1950s. Alkali halides and especially LiF doped with various other atoms are still the most popular phosphors. In the following we will try to describe the mechanism of TL in alkali halides, as LiF is the material most commonly used to build TL dosimeters.

The crystalline structure of an alkali halide (such as LiF or CaF) consists of two interpenetrating cubic lattices of alkali and halogen ions as illustrated in figure 1.6. There are two general categories of lattice defects: (i) intrinsic defects (caused by thermal vibrations of the ion) and (ii) extrinsic defects (produced by substitution of an ion with a different one).

The population of intrinsic defects is strongly dependent on the temperature of the material, given that they are produced due to the thermal movement of the atoms, which is more intense at higher temperatures. The higher the temperature, the greater the number of such defects in a crystal. Moreover, the number of such defects in a crystal exhibits inertia in abrupt temperature changes. We may thus

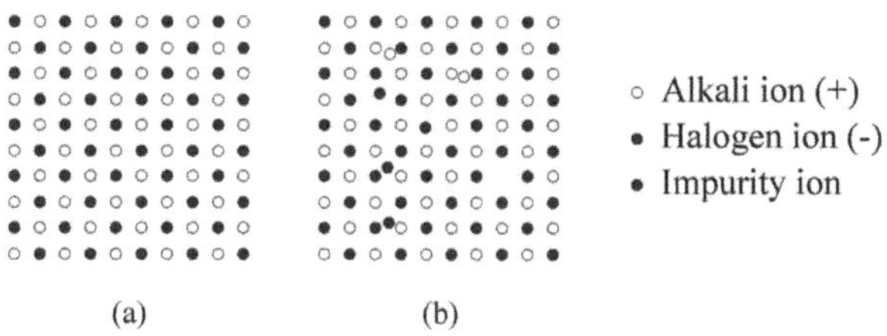

Figure 1.6. Perfect alkali halide ionic structure (not real) (a); imperfect crystal containing defects of various types (real) (b).

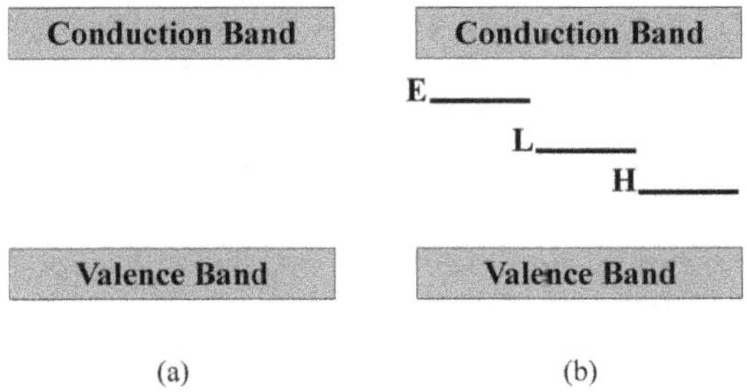

Figure 1.7. Energy band diagrams. Ideal perfect crystal (a) and real imperfect crystal (b).

maintain the number of defects to a high temperature by simply decreasing abruptly the temperature of the crystal. The term 'defect' describes all the specific areas of the crystalline structure that differ from the ideal one. Thus, a missing ion (halide or halogen) from a lattice location, the existence of an ion at an intermediate location, and the substitution of an ion with an impurity ion are some simple defects, whereas the combination of simple defects yields complex defects.

In the following text we explain why the presence of defects in the crystalline structure of the crystals is crucial in TL (Kumar *et al* 2007). Actually, there would be no TL effect if the lattices of the crystals were perfect. The energy band structure for an ideal crystal may be represented by an energy band diagram as shown in figure 1.7(a). The valence band includes all the energy states of an electron corresponding to bound states and the conduction band to all the energy states corresponding to unbound states. Between the valence band and conduction band there is the forbidden energy gap, where no energy states are allowed. All electrons in the conduction band are free to move through the crystal lattice. Without the influence of external forces, it is highly improbable for an electron from the valence band to cross the forbidden energy gap reaching the conduction band, however, if this happens, we finally have an electron in the conduction band (free negative charge) and a positive ion in the lattice corresponding to a free positive charge generally termed a hole. In a real crystal containing defects of simple and/or complex nature, energy levels are allowed in the forbidden energy gap as shown in figure 1.7(b), due to the existence of these defects.

For instance, a negative ion vacancy in the lattice (missing halogen) is a region of excess positive charge and so it is regarded as a potential trap for a free electron corresponding to an energy level E. Similarly, a positive ion vacancy (missing halide) is a region of excess negative charge and so it is regarded as a potential trap for free positive charges (holes) corresponding to an energy level H. In the following description of a simplified model of TL we suppose that energy level E represents an electron trap, energy level H represents a hole trap and L is a luminescence center where electrons and holes recombine with photon emission.

1.2.4 Production of thermoluminescence by ionizing radiation

When ionizing radiation is absorbed by a thermoluminescent material free electrons are produced. This is equivalent to transferring electrons from the valence band to the conduction band. These electrons are now free to move within the crystal lattice. If trapping levels such as *E* (figure 1.6(b)) are present, the liberated electrons may become trapped. The production of free electrons is associated with the production of free positive holes, which may also move around in the crystal lattice and become trapped in trapping levels such as *H*. The trapped electrons remain in their traps provided that they do not acquire sufficient energy to escape (Horowitz and Datz 2011, Horowitz and Moscovitch 2013).

The escape of electrons from their traps is related to the depth of the traps and the temperature of the material. The depth of an electron trap is defined as the energy distance between the trapping energy level and the lower energy level of the conduction band (ΔE in figure 1.5). If we heat the material, the temperature of the material is raised and trapped electrons may get sufficient thermal energy to be released (Horowitz 1981). Released electrons may recombine with holes at luminescence centers such as *L* and the excess in energy is irradiated as optical photons. Although electron capture and heat-boosted recombination with a hole is the mechanism of TL, other electron–hole recombination processes are possible such as fluoroscopy caused by immediate electron–hole recombination, delayed or immediate electron–hole recombination with thermal degradation of the excess in energy without emission of the optical photons. The production of TL in a material by exposure to ionizing radiation includes three steps: (i) ionization of the material and electron trapping of the liberated electrons, (ii) heating the material (making captured electrons escape from their traps), and (iii) electron and hole recombination with photon emission.

1.2.5 The glow curve

A material containing defects of only one type gives rise to electron traps that correspond to the same energy level. In other words, the energy depth of all the traps is the same, that is ΔE below the bottom of the conduction band. If we suppose that at time *t*, the temperature of the material is *T* (K) and *n* electrons are trapped, the probability *P* of release of a single electron per unit time is given by the Boltzmann's equation (1.1). If the material is heated at a uniform rate $R = dT/dt$, the glow intensity (*I*) emitted by the material is given by the following equation:

$$I = n_0 Cse^{-\int (1/R)s\,\exp(-\Delta E/kT)dT} e^{-\Delta E/kT} \tag{1.2}$$

where n_0 is the number of electrons present in the trap at time t_0 and temperature T_0, and *C* is a constant related to luminescence efficiency, that is the ratio of recombined holes to electrons that give rise to optical photons. The plot of *I* versus *T* is called the glow curve. Initially the luminescence (*I*) rises exponentially as electrons are ejected from the traps by the increasing temperature and after reaching a maximum (glow peak) it falls to zero as the store of trapped electrons is depleted. The greater the energy depth ΔE of the trapping level the higher the temperature that the glow peak maximum occurs. The glow peak moves at higher temperatures as the heating rate *R* is increased. The area under the curve is directly proportional to the number of electrons initially trapped. The traps still exist after they have been emptied allowing the TLD to be used again.

The theoretical model discussed so far refers to the case of a crystal presenting only one trapping energy level. In real thermoluminescent materials many different trapping levels are present, each one associated with one simple or complex lattice defect (Duggan and Kron 1999). Each trapping level gives rise to a glow peak maximum. The area under the peak and the peak height of each glow peak depends on the number of the associated electron traps. This in turn depends on the lattice defects corresponding to these electron traps. Finally, the number of lattice defects depends on the type and amount of impurity atoms present as well as on the thermal history and treatment of the material.

Lithium fluoride doped with magnesium (~300 ppm) and titanium (~15 ppm), LiF:Mg:Ti, is the most commonly used and the most widely researched thermoluminescent material. All the research on this phosphor has led to the conclusion that the TL process in LiF:Mg:Ti is very complicated and is critically dependent on: (a) the amount, type, chemical form and method of introduction into the lattice of the impurities present and (b) the thermal, optical and mechanical treatment of the phosphor during its manufacture and use.

The glow curve of LiF:Mg:Ti contains a large number of glow peaks (Weizmann *et al* 1999). When, for instance, it is heated at 400 °C and cooled quickly to normal ambient temperature (~20 °C), the resulting glow curve after irradiation contains at least six glow peaks between normal ambient temperatures and 300 °C (figure 1.8). These are called peaks 1 (~60 °C), 2 (~120 °C), 3 (~170 °C), 4 (~190 °C), 5 (~210 °C) and 6 (~285 °C). The precise read-out temperature and the resolution of each peak depends on the heating rate employed during read-out. Peaks 4 and 5 are very close so it is difficult to discriminate. They both contribute (mainly 5) to the peak normally used for practical dosimetry. The low-temperature peaks exhibit high fading of the stored signal even at normal ambient temperature (escape of electrons from their traps before heating). It is possible, however, to reduce effectively the height of the low-temperature peaks by decreasing the number of associated electron

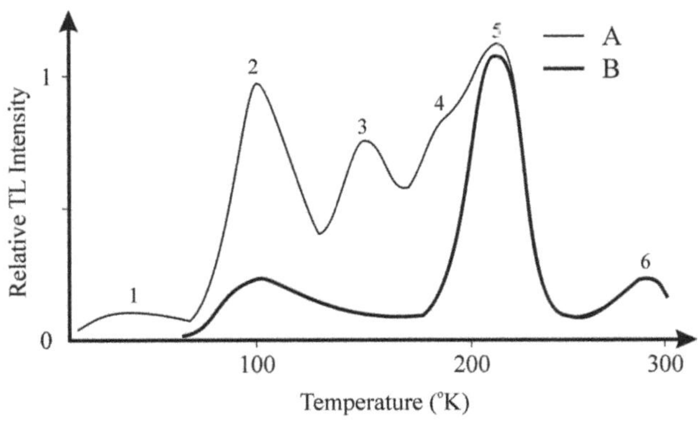

Figure 1.8. Glow curves of LiF:Mg:Ti. Glow curves of LiF:Mg:Ti when it is annealed for 1 h at 400 °C followed by: A cooling (1000 °C min^{-1}) to normal ambient temperature and B annealing for 16 h at 80 °C, before irradiation.

traps. This is achieved by thermally annealing the material before the irradiation for 1–2 h at 100 °C or 16–24 h at 80 °C after the 1 h anneal at 400 °C (Stadtmann *et al* 2006). This procedure results in a much more satisfactory glow curve (B in figure 1.8).

1.2.6 The role of impurity atoms Mg,Ti in the TL process of LiF:Mg:Ti

Magnesium and titanium atoms play an important role in the TL process of LiF: Mg:Ti, as it occurs with the impurities of every phosphor (Bilski *et al* 1996, Delgado 1996, Piters and Bos 1995). The magnesium ions are presumed to form electron traps in combination with certain defects of the lattice. The role of titanium ions is much more unclear although it is believed to be associated with the formation of luminescence recombination centers.

Magnesium

The substitution of the monovalent lithium ion (Li^+) by the divalent magnesium ion (Mg^{++}) results in an excess positive charge at that lattice site. Coulombic attraction to lithium ion vacancies results in the formation of dipoles consisting of a substitutional Mg ion in combination with a Li vacancy. These dipoles are believed to aggregate, under certain thermal conditions, forming dimers, trimers and higher order complexes. There is evidence that the simple dipole arrangement is associated with the electron traps responsible for the low-temperature peaks 2 and 3 (figure 1.8). Dipole complexes are associated with the main dosimetry glow peaks 4 and 5. The aggregation of simple dipoles to form complexes is critically dependent on temperature and heating/cooling rates. This mechanism of trap formation can be used to explain the change in the relative heights of the glow peaks due to pre-irradiation annealing of the phosphor. As mentioned above, the standard pre-irradiation thermal annealing of the UP phosphor at 400 °C for 1 h followed by cooling to normal room temperature produces a glow curve with six peaks within the range between normal ambient temperature and 300 °C, as shown in figure 1.8. This is valid for any annealing temperature above 180 °C, but as the annealing temperature gets higher, complexes of dipoles are broken down to simple dipoles producing relatively high 2 and 3 peaks. If the phosphor after 1 h anneal at 400 °C is annealed at 100 °C for 1–2 h, aggregation of the dipoles occurs enhancing the main dosimetry peaks (4 and 5) at the expense of the low-temperature peaks. This is more intense if instead of the 1–2 h annealing at 100 °C we use 16–24 h at 80 °C.

The model of aggregation of dipoles may be used to explain the effect of the cooling rate from 400 °C. If the phosphor is cooled rapidly, a number of electron traps associated with simple dipoles are 'frozen' into the lattice resulting in a relative enhancement of peak 2. A slower cooling allows aggregation that relatively enhances peaks 4 and 5.

Titanium

All crystalline materials are grown under air from the melt and contain hydroxyl OH complex ions in concentrations of several tens of ppm. The TL efficiency of LiF:Mg: Ti has been observed to increase with increasing concentration of titanium ions presenting a maximum at 7 ppm, while in phosphors with higher concentrations in titanium the TL sensitivity seems to be controlled by OH ion concentration and vice versa. Titanium may also be present in the divalent state (Ti^{++}) possibly forming dipoles with Li vacancies in a similar manner to Mg. So titanium concentration affects the relative shape of the glow peaks.

1.2.7 Fading

As mentioned previously, the glow curve B in figure 1.8 is much more satisfactory than A. The reason is fading. When a thermoluminescent material is exposed to ionizing radiation the absorbed dose is recorded by measuring the glow intensity, which is proportional to the number of electrons that remain trapped. However, electrons trapped in shallow traps (which are associated with low-temperature peaks in figure 1.8) may acquire sufficient energy to be released spontaneously, even at normal ambient temperature. The decrease of trapped electrons is more intense the more time elapsing after the irradiation and before the measurement of the material. This is definitely highly irritating when thermoluminescent materials are used in dosimetry because fading makes the measurement depend on the time elapsing after the irradiation until the measurement and this is undesirable.

Thermal fading

The release of electrons from trapping centers is governed by Boltzmann's equation (1.1). The half-life $t_{1/2}$ of a particular trapping center is defined as the time needed for the number of trapped electrons to fall to half. The half-life $t_{1/2}$ is given by the following equation:

$$t_{1/2} = \ln(2/s)e^{(\Delta E/kT)} \tag{1.3}$$

The above equation demonstrates that $t_{1/2}$ increases exponentially as the energy depth of the trap ΔE is increased and decreases as temperature T increases.

The choice of glow peak five (figure 1.8) for the measure of the TL signal is mainly based on the fact that this peak exhibits a half-life of the order of many years, thus the fading effect is not significant. However, the fading effect is significant for the low-temperature peaks (Vergara Saez *et al* 1999). Thus, to minimize the error due to fading we try to get rid of low-temperature peaks. This is done by two means:

(a) As mentioned above, a 16–24 h anneal at 80 °C after the basic 1 h at 400 °C anneal leads to the enhancement of the main dosimetry peaks 4–5 at the expense of the low-temperature peaks 2–3. Alternatively, a 1–2 h at 100 °C anneal may be used which, however, results in less degradation of the low-temperature glow peaks.

(b) A post-irradiation pre-measurement anneal can also be used to fade intentionally the low-temperature peaks before read-out of the thermoluminescence signal from the main dosimetry peak 5. This procedure is used in combination to the second annealing protocol discussed above (1 h at 400 °C plus 1 h at 100 °C).

Optical fading

Just as energy in the form of heat induces the depopulation of a TL trap so can energy in the form of a photon. If an irradiated phosphor is exposed to photons of visible and especially ultraviolet (UV) range, trapped electrons may absorb such photons and be ejected from their traps or redistributed in other trap energy levels. This is an additional cause of a loss of signal. It can be avoided by simply storing the phosphors in opaque containers during the time elapsing from the irradiation until the measurement. Similarly, just as energy in the form of ionizing radiation can cause a TL trap to be populated, so can energy in the form of a photon. Prolonged exposure of unirradiated phosphors to the UV component of natural or artificial light can also cause a spurious TL signal. Therefore, storing of phosphors in opaque containers is recommended at all times.

1.2.8 Linearity/supralinearity and sensitization

For the thermoluminescent signal we consider the height of peak 4–5 or the integral under the peak 4–5 of the glow curve, with the integral being the most popular choice. Obviously, in both cases, the TL signal is strictly proportional to the emitted light intensity.

Linearity

The plot of the TL signal against the absorbed dose D gives the dose-response curve for the irradiated material. The TL absorbed dose response is an important characteristic of a thermoluminescent phosphor. At relatively low values of absorbed dose (usually up to 10 Gy) the response is linear:

$$\mathrm{TL} = cD + B \tag{1.4}$$

where TL is the thermoluminescent signal for an absorbed dose D, c is the TL emitted per unit absorbed dose and B is the background signal from an unirradiated phosphor (zero dose signal).

Supralinearity

Above 10 Gy the response is supralinear, saturating at ~1000 Gy and then falling off rapidly (figure 1.9). The supralinear behavior of a phosphor may appear to be an appealing characteristic as it yields increased sensitivity, but in reality is generally undesirable as extra calibration is required when the material is used for high dose measurements (Olko *et al* 2001).

An absorbed dose at approximately 80% of 1000 Gy is generally taken to be the practical upper absorbed dose limit. The detection threshold is usually considered to be three times the standard deviation of the values of the background of a reasonably sized (~50) sample of phosphors.

Sensitization

When LiF:Mg:Ti is irradiated in high dose rates (300–1000 Gy) and then it is annealed at ~300 °C for a short period (~10 min) the TL efficiency is increased by a

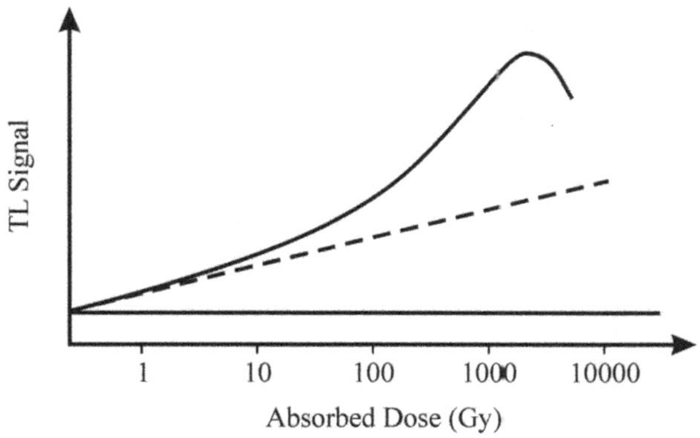

Figure 1.9. A typical thermoluminescence signal absorbed dose-response curve.

factor of up to five. This procedure is used in some special cases when high sensitivity is required and it is called sensitization. A 2 h anneal at 360 °C may remove this enhancement.

1.2.9 Photon energy response of thermoluminescent materials

For relatively low doses the TL signal is proportional to the total absorbed dose. The absorbed dose to the phosphor is related to the exposure and the mass energy absorption coefficient of the material:

$$D_{\mathrm{phosphor}} = cX(\mu_{\mathrm{en}}/\rho)_{\mathrm{phosphor}}/(\mu_{\mathrm{en}}/\rho)_{\mathrm{air}} \tag{1.5}$$

where D_{phosphor} is the dose absorbed by the phosphor in Gy, c is a constant that equals the average energy required to produce an ion pair in air independently of the photon energy ($C = W_{\mathrm{air}}/e$), X is the exposure in C/kg and $(\mu_{\mathrm{en}}/\rho)_{\mathrm{phosphor}}$, $(\mu_{\mathrm{en}}/\rho)_{\mathrm{air}}$ are the mass absorption coefficients for phosphor and air, which depend on photon energy. The mass energy absorption coefficient for a certain thermoluminescent material that consists of many atoms of the basic lattice plus few doping atoms, may be calculated from the formula:

$$(\mu_{\mathrm{en}}/\rho)_{\mathrm{phosphor}} = \sum(\mu_{\mathrm{en}}/\rho)_i w_i \tag{1.6}$$

where $(\mu_{\mathrm{en}}/\rho)_i$ is the mass energy absorption coefficient of the ith elemental constituent of the phosphor and w_i is the fraction of that element in the phosphor.

The quantity $F = \{c(\mu_{\mathrm{en}}/\rho)\}_{\mathrm{phosphor}}/(\mu_{\mathrm{en}}/\rho)_{\mathrm{air}}\}$ is a function of photon energy and is very important in TL dosimetry. It is commonly called the roentgen-to-gray (R to Gy) conversion factor and when it is multiplied by exposure X (R) it yields dose D (Gy).

1.2.10 Calibration of thermoluminescent dosimeters

TLDs do not provide an absolute measure of a radiation absorbed dose. This means that they must be calibrated against a primary measurement system either directly or, most commonly, through a calibrated secondary system such as an ionization chamber. For a specific photon energy E and relatively low absorbed doses the TL sensitivity calibration factor C of a TLD is given by:

$$C(E) = \mathrm{TL}/D_{\mathrm{phosphor}} \tag{1.7}$$

where TL is the measured TL signal resulted from the absorption by the TLD of dose D_{phosphor}. This expression comes from equation (1.4) by subtracting the background signal from the total. The calibration factor C of a TLD is calculated using equation (1.7) by exposing an ionization chamber along with the TLD to a photon beam with a specific energy. The produced TL signal is measured using a TLD reader and the dose D is estimated from the exposure X of the beam used to irradiate the TLD. The dose is estimated from equation (1.8):

$$D_{\mathrm{phosphor}} = FX \tag{1.8}$$

where F is the R to Gy conversion factor appropriate to the photon energy of the beam and the specific phosphor material.

A Co^{60} source with photon energies 1.17 and 1.33 MeV (mean photon energy 1.25 MeV) and half-life $T_{1/2} = 5.3$ years, is often used for calibration of TLDs. Due to the high energy photon beam, the dosimeter and the ionization chamber must be surrounded by sufficient build-up material to ensure electronic equilibrium. Once the calibration coefficient C for a TLD has been calculated for a given photon energy (for instance 1.25 MeV), the calibration coefficient for any other photon energy may be expressed as a fraction or a multiple of C. This can be done by means of the relative energy response (RER_E):

$$\mathrm{RER}_E = [(\mu_{\mathrm{en}}/\rho)_{\mathrm{phosphor}}/(\mu_{\mathrm{en}}/\rho)_{\mathrm{air}}]_E/[(\mu_{en}/\rho)_{\mathrm{phosphor}}/(\mu_{\mathrm{en}}/\rho)_{\mathrm{air}}]_{1.25\,\mathrm{MeV}} \tag{1.9}$$

The RER of several TL materials is plotted as a function of photon energy in figure 1.10.

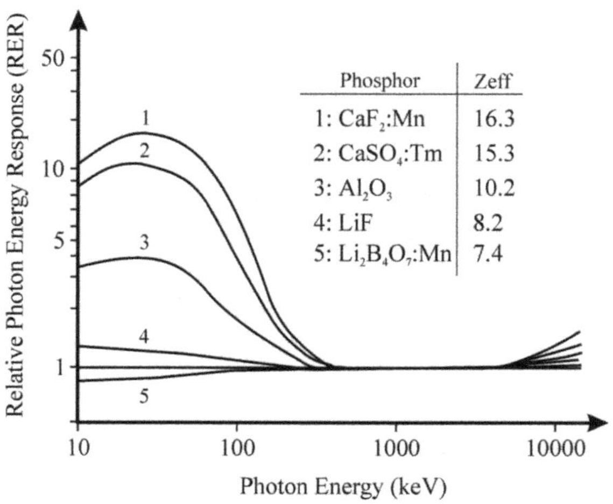

Figure 1.10. Theoretical photon energy responses of some thermoluminescent materials with respect to air (1.00). The effective atomic number of each material is also shown. Adapted from McKinlay (1981).

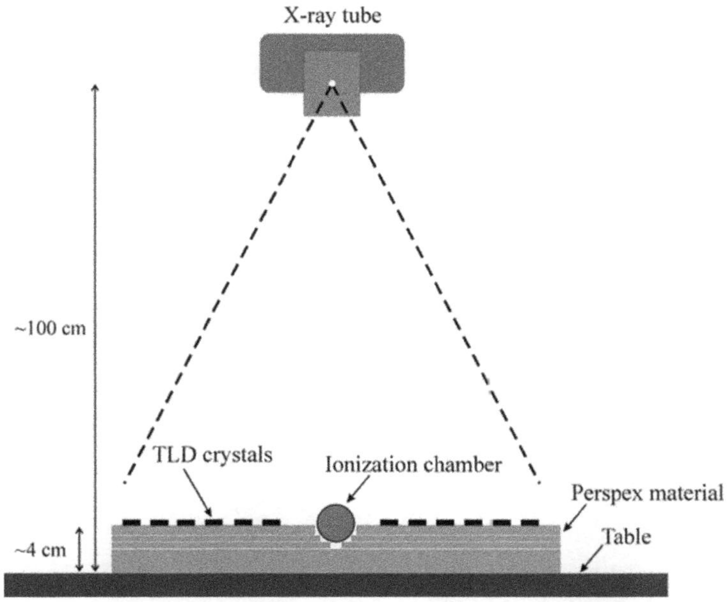

Figure 1.11. A typical configuration set-up employed to calibrate TLD crystals against a range of known absorbed doses (Gy) using a clinically available radiographic unit.

The difficulty in calibrating TLDs at a specific photon energy and estimating the relative energy response of the TLDs at the photon energy used in actual measurements suggests that it is preferable to calibrate the TLDs using a radiation source at the same photon energy as that used in actual measurements. A typical irradiation set-up that might be employed to calibrate TLD crystals using a radiographic unit is illustrated in figure 1.11.

1.2.11 TLD reader

The purpose of a TLD reader is to measure the output of thermoluminescent dosimeters after their irradiation. The reader must have the capability to controllably and reproducibly heat the TL dosimeters, collect the emitted light and measure the light output. The main components of a commonly used TLD reader (figure 1.12(a)) are the TL dosimeter heating system, the light collection system, the signal measuring system, and (*d*) display system. A commercially available TLD reader device is shown in figure 1.12(b).

Heating system

The primary purpose of the heating system in a TLD reader is to heat the TL sample (dosimeter) until the electrons are liberated from the dosimetry traps. In addition to this, it is sometimes used to fade rapidly the low-temperature glow peaks, by means of a pre-read anneal at low temperature. In cases when TLDs are used to measure high doses, TLD readers may be used to offer a high-temperature anneal, although usually this is performed by dedicated annealing ovens.

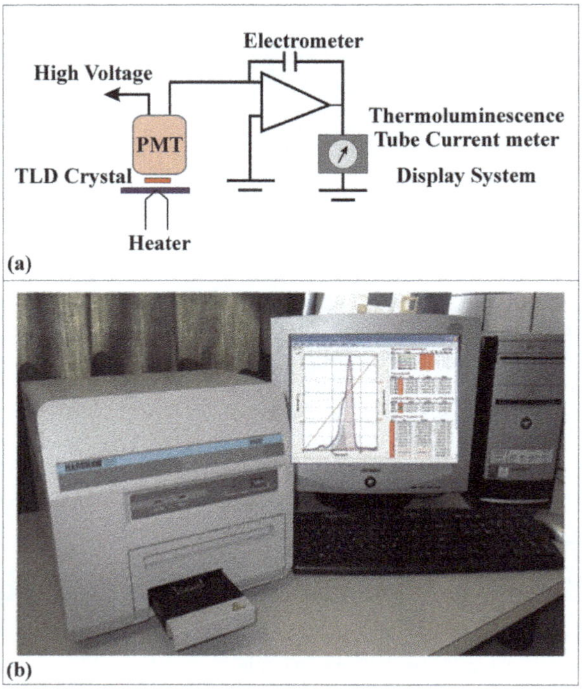

Figure 1.12. (a) The configuration set-up of a typical TLD reader, (b) a commercially available TLD reader device. Courtesy of the UoC.

Light collection system

The light collection system is responsible for the efficient collection of TL emitted from the heated TL dosimeters, rejecting if possible any other optical radiation (such as infrared radiation emitted from the heated tray, etc) and the conversion of the collected light to electrical signal. In all TLD readers the light collection system is based on a photomultiplier tube (PMT). PMTs make use of the phenomena of photoemission and secondary electron emission to detect very low light intensities. Photoemission is the ejection of electrons from a material as a result of exposure to optical radiation. The electrons emitted from the photocathode by incident light are accelerated and focused onto another emission surface. The PM tube acts as an electron multiplier, with 1–100 million electrons collected at the final electrode for each electron released from the photocathode. For a specific photon spectrum, the response of the phototube at any given high voltage is linear. Many materials used to make the photocathode have their peak sensitivity around 400 nm in the same region as the TL emitted from the most commonly used phosphors. To provide good rejection of the unwanted infrared (thermal) radiation originating from the tray as the temperature is raised, a cut-off high pass filter is often used. The photocathode is very sensitive to exposure to room daylight. Therefore, it should operate under complete darkness. However, even under conditions of complete darkness some electrons are ejected from the photocathode due to thermionic emission. The current produced by these electrons is called dark current. Dark current may be reduced by keeping the temperature of the photocathode low, but never eliminated completely.

Signal measuring system
The photomultiplier small current, which is proportional to the TL luminous flux, is measured by an electrometer. The electrometer amplifies this current. The TL signal may finally be represented by the integrated charge collected by a series of capacitors.

1.2.12 Thermoluminescent materials

The characteristics of the most commonly used thermoluminescent materials are listed in table 1.4.

1.2.13 The physical properties of LiF

Lithium fluoride is the most commonly used thermoluminescent material (especially LiF:Mg:Ti). It is an alkali halide with density $\rho = 2.64\ \rho H_2O$ and effective atomic number $Z_{eff} = 8.2$. It has acceptable resistance to most chemicals and is practically insoluble in water. The standard material (named TLD-100) produced for the first time by the Harshaw Chemical Co (Cleveland, USA) in 19 605, contains the natural isotopic composition of Li (7.5% Li-6 and 92.5% Li-7). The material is also available with different isotopic constituents that is Li-6 enriched (TLD-600) and Li-7 depleted (TLD-700). Both lithium isotopes (Li-6 and Li-7) are sensitive to gamma and beta radiation. Li-7 is practically insensitive to neutron radiation, in contrast to Li-6, which is very sensitive to neutron radiation. Therefore, TLD-100, 600, 700 are generally reported to have similar sensitivities to gamma and beta radiation but different sensitivities to thermal neutrons, so they can be used in combination in a mixed gamma-neutron radiation field.

Production of the material – physical forms of LiF dosimeters
TLD-100, 600 or 700 is produced by homogenous melting of lithium fluoride, magnesium fluoride, lithium cryolite and lithium titanium fluoride resulting in a phosphor containing 300 ppm magnesium and 10–20 ppm titanium. A single crystal is solidified from the melt and then pulverized. Extruded ribbon dosimeters are formed by compressing the original mixture at an elevated temperature causing it to extrude through an aperture. The extrusion is then cut into pieces that are polished.

Li:Mg:Ti is available in a wide range of forms including powder, single crystals, extruded ribbons and micro-rods. Single crystals of TL material, while providing a very sensitive form of dosimeter, display a very non-uniform distribution of TL sensitivities. It has been found that by grinding up many such crystals and mixing the resultant powder, a reasonably uniform TL sensitivity can be achieved. Although there are problems in using powder dosimeters, such as loss of TL material, different TL sensitivity according to grain size, uncertainty in the uniformity of distribution and positioning of the powder on the reader heating tray etc, LiF:Mn:Ti in the form of powder is still used in many cases. The extruded and hot pressed forms of

Table 1.4. The characteristics of the most commonly used thermoluminescent materials.

	TLD-100, LIF:Mg, Ti (lithium fluoride)	TLD-200, CaF_2:Dy (calcium fluoride dysprosium)	TLD-300, CaF_2:Tm (calcium fluoride thulium)
Applications	Health and medical physics dosimetry	Environmental dosimetry	Environmental dosimetry, fast neutron dosimetry
Forms	Chips, rods, powder, cards	Chips, rods, powder, cards	Rods, chips, single crystals
Density	2.64 g cm^{-3}	3.18 g cm^{-3}	
Z_{eff}	8.2	16.3	
Temp. of main glow curve peak	195 °C	180 °C	150 °C, 240 °C
Sensitivity at ^{60}Co relative to LIF	1	30	3
Energy response (30 keV/^{60}Co)	1.25	12.5	
Useful range	1 μR–10^6 R	10 μR–10^6 R	
Light (UV) sensitivity	Very low	High	
Supralinearity	>100 R	–	
Fading	5%/year at 20 °C	5% in 50 days	
Anneal	Pre-irradiation: 400 °C at 1 h + (100 °C at 2 h or 80 °C at 16 h). Post-irradiation: 100 °C at 10 min	Pre-irradiation: 500 °C at 1 h. Post-irradiation: 100 °C at 20 min	400 °C at 1 h
	TLD-400, CaF_2:Mn (calcium fluoride manganese)	TLD-500, Al_2O_3:C (aluminum oxide carbon)	TLD-600, 6LIF:Mg, Ti (lithium-6 fluoride)
Applications	Environmental and high dose dosimetry	Environmental dosimetry	Neutron dosimetry
Forms	Chips, rods, powder, cards, bulbs	Disks	Chips, rods, powder, cards
Density	3.18 g cm^{-3}	2.7 g cm^{-3}	2.64 g cm^{-3}
Z_{eff}	16.3	10.2	8.2
Temp. of main glow curve peak	260 °C	209 °C	195 °C
Sensitivity at 60Co relative to LIF	1.7–13	30	1
Energy response (30 keV/^{60}Co)	13	2.9	1.25

(Continued)

Table 1.4. (*Continued*)

	TLD-100, LIF:Mg, Ti (lithium fluoride)	TLD-200, CaF_2:Dy (calcium fluoride dysprosium)	TLD-300, CaF_2:Tm (calcium fluoride thulium)
Useful range	100 μR–3 × 10^5 R	50 μR–100 R High rate dependence	mR–10^5 R
Light (UV) sensitivity	High	High	
Supralinearity		–	
Fading	10% in 1st 24 h, 15% total in 2 weeks	5% per year	5% per year at 200 C
Anneal	Pre-irradiation: 500 °C at 1 h. Post-irradiation: 100 °C at 10 min	Pre-irradiation: 800 °C at 15 min	Pre-irradiation: 400 °C at 1 h + (100 °C at 2 h or 80 °C at 16 h). Post-irradiation: 100 °C at 10 min
	TLD-700, ^{7}LIF:Mg, Ti (lithium-7 fluoride)	TLD-700H, ^{7}LIF:Mg, Cu, P (lithium-7 fluoride copper)	TLD-800, $Li_4B_4O_7$:Mn (lithium borate manganese)
Applications	Gamma dosimetry	Environmental dosimetry	High dose dosimetry, neutron dosimetry
Forms	Chips, rods, powder, cards	Chips, powder, cards	Chips, powder, cards
Density	2.64 g cm^{-3}	2.08 g cm^{-3}	2.4 g cm^{-3}
Z_{eff}	8.2	–	7.4
Temp. of main glow curve peak	195 °C	195 °C	200 °C
Sensitivity at ^{60}Co relative to LIF	1	10	0.15
Energy response (30 keV/^{60}Co)	1.25	1.25	0.9
Useful range	1 mR–3 × 10^5 R	1 mR–3 × 10^5 R	50 mR-10^6 R
Light (UV) sensitivity			
Supralinearity		–	
Fading	5%/year at 20 °C	Negligible	<5% in 3 months
Anneal	Pre-irradiation: 400 °C at 1 h + (100 °C at 2 h or 80 °C at 16 h). Post-irradiation: 100 °C at 10 min	Pre-irradiation: 240 °C at 10 min. Post-irradiation: 165 °C at 10 s (reader preheat)	Pre-irradiation: 300 °C at 15 min

dosimeter are solid and available in two main styles: the ribbon (commonly called chip) and the rod (commonly called micro-rod). They present similar TL characteristics with powder dosimeters with the advantage that they are solid so can be easily handled and washed. A variety of sizes are available but currently the most popular ones are $3 \times 3 \times 0.9$ mm chips and $1 \times 1 \times 6$ mm micro-rods.

Spectral emission
The TL output spectrum for Li:Mg:Ti has its peak at 400 nm (characteristic blue), corresponding to the optimum PMT sensitivity, resulting in an increased overall response.

Glow curve and thermal treatment of the material
There are two basic lithium fluoride glow curve shapes as illustrated in figure 1.8 named quenched (A) and annealed (B) (Muniz *et al* 1996, Horowitz and Moscovitch 1986, Horowitz *et al* 1993). The terms 'quenched' and 'annealed' refer to the pre-irradiation thermal treatment of the material (the heating procedure chosen). 'Quenched' describes a rapid heat up and cool down, whereas 'annealed' describes a very slow heat up and cool down. Annealing can only be achieved by heating in an oven as TLD readers always quench TLDs. An annealed glow curve (B in figure 1.8) is generally preferable. However, there are an infinite variety of glow curves between (A) and (B). The shape of the curve depends on the heating process being used. Therefore, it is imperative that once a heat treatment method is selected, it is done exactly the same all of the time. The recommended and most commonly used annealing protocols are: (a) 1 h at 400 °C followed by 2 h at 100 °C + pre-measurement anneal at 100 °C for 10 min, and (b) 1 h at 400 °C followed by 20 h at 80 °C.

Fading
The peak of the Li:Mg:Ti glow curve generally used in dosimetry is 5. The half-life of the trapping centers associated with peak 5 is 80 years so the loss of stored signal, associated with peak 5, is less than 5% at normal ambient temperature, even when the elapsing time between irradiation and measurement is one year. We may almost eliminate the error introduced due to fading by measuring the TL samples immediately after irradiation or keeping the irradiated phosphors at low temperature and complete darkness until they are measured.

Absorbed dose response
The response of a lithium fluoride TLD is very nearly linear up to 4 Gy and supralinear above that value (figure 1.9). The exact onset and degree of supralinearity depends on the linear energy transfer of the radiation measured, moving to higher doses as photon energy is decreased. The lowest detectable absorbed dose depends on the physical form of the dosimeter, the read-out equipment and the user technique. For chip, micro-rod and powder dosimeters it usually ranges from 50 to 100 μG. The upper dose limit is 1000 Gy, while doses greater than 10^4 Gy cause permanent radiation damage in the lattice and an irreversible loss of sensitivity. The

response of LiF:Mg:Ti has been shown to be independent of the absorbed dose rate up to 10^8 Gy s^{-1}.

Tissue equivalence
The effective atomic number of LiF is $Z_{LiF} = 8.2$, while for tissue it is $Z_{tissue} = 7.4$. For most applications LiF is approximately tissue equivalent. LiF has a reasonably flat photon energy response, as shown in figure 1.10, with a maximum over-read of 1.3 at 20–30 keV. This means that when LiF TLDs are calibrated in 1.25 MeV photon energy beam (^{60}Co) and used to measure the radiation of photon energy in the diagnostic range there is an overestimation of the absorbed dose up to 30%. This can be avoided if the dosimeters are calibrated at the same energy photon beam as the one to be measured.

1.2.14 Sources of error in thermoluminescence dosimetry

Thermoluminescence dosimetry is a rather complex method of measuring radiation doses, therefore it involves in practice many sources of errors. There are random errors associated with the random nature of radiation production and absorption, which are related to the precision (reproducibility) of the measuring method. There are also systematic errors that are much more irritating and are related to the accuracy of the measuring method. Considerable care must be taken in order to minimize all systematic as well as random errors. The three main components of an established TL dosimetry technique are the TLDs, the TLD reader and the pre- and post-irradiation thermal treatment of the dosimeters. Errors may be introduced in any of these components.

Errors associated with the type of TLDs

(a) Variation of TL sensitivity within a batch of dosimeters: In part this is inevitable as it depends on the production process of the TLD material. However, the additional TLD sensitivity variation, produced by the loss of phosphor and change of the optical properties of the material during usage, may be significantly decreased by careful handling of the material.
(b) Variation of TL response with radiation energy: This is an inherent characteristic of the material depending on the variation of mass absorption coefficient of the TL material with photon energy. The choice of the TLD size (dimensions of the phosphors used) in an application is important in order to eliminate the effects of self-shielding or lack of electronic equilibrium. Also, the TLD manufacturer must define the useful energy range of the dosimeters.
(c) Loss of signal due to fading: Errors due to the fading effect can be significantly reduced by choosing a TL material with deep traps and certain thermal treatment protocols before, during and after read-out.
(d) Sensitivity to light: As TL materials are sensitive to daylight (light absorption reduces the TL signal in irradiated phosphors and produces

signals in unirradiated phosphors), TLDs should be kept in opaque containers.

Errors associated with the TLD reader

(a) Thermal contact between the heater and detector: Reproducible thermal contact between the heating element and TLDs is essential as when it is varied the thermal treatment of the material is not reproducible.
(b) Photomultiplier tube efficiency: The light collection system efficiency must be constant. The PMT window should be kept clean and the positioning of the TLDs during read-out should be repeated exactly.
(c) Gain of the ND reader electronics: The gain of the electronics depends (among other things) on ambient temperature. Before starting to measure TLDs, it is a good idea to run three or four read-out cycles with the standard light source. This way, the gain stability is checked daily and at the same time warm-up of the device is achieved.
(d) Background readings: Increased background readings affect the precision of the method especially in low dose measurements. Keeping TLDs and the heating tray clean as well as introducing nitrogen gas during read-out to eliminate oxygen effect may reduce the background significantly.

Errors associated with thermal treatment of the material

(a) Heating cycle: The performance of the heating cycle in an absolutely reproducible manner is of major importance. The cooling rate after read-out has to be the same always as it strongly influences the TL sensitivity.
(b) Thermal annealing: Whatever the procedure adopted it should be carried out in a highly reproducible manner. Furthermore, a prolonged storage period after annealing and before irradiation may also introduce discrepancies.

1.3 Optically stimulated luminescence dosimeters (OSLDs)

1.3.1 The fundamentals of optically stimulated luminescence (OSL) dosimetry

The operation principle of OSL dosimetry is similar to TLDs, except that the read-out process includes illumination of the dosimeter material instead of heating (Akselrod and Gorelova 1993, Evans *et al* 1994). In OSL dosimetry, optical transitions can occur when the detector material is exposed to light. The energy levels in the crystalline structure of the dosimeter consist of delocalized energy bands; the valence and conduction bands, which are created by the periodic potential of the crystalline material (figure 1.13). The band gap that separates the valence from the conduction band contains localized energy levels, which are produced by defects in the structure of the crystal material. When the crystal is exposed to ionizing radiation, electrons are excited to the conduction band, producing holes in the valence band. The electron–hole pairs move freely in the crystal lattice until they recombine with each other or are captured by the localized energy levels, which act as traps. The trapped charge concentration at these localized energy levels gives a record of the radiation dose absorbed by the crystal material. This record can be read by stimulating the trapped

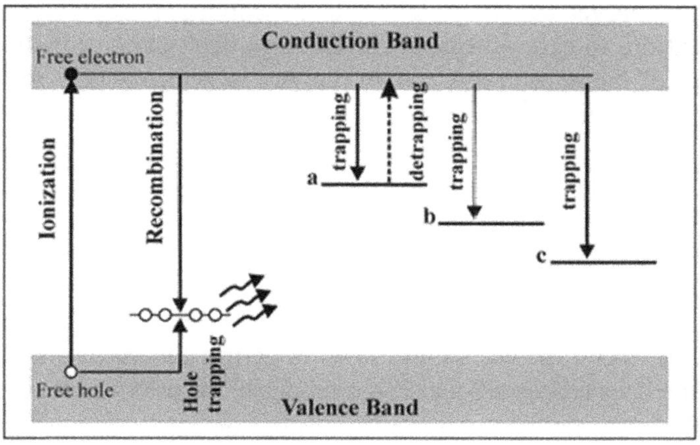

a

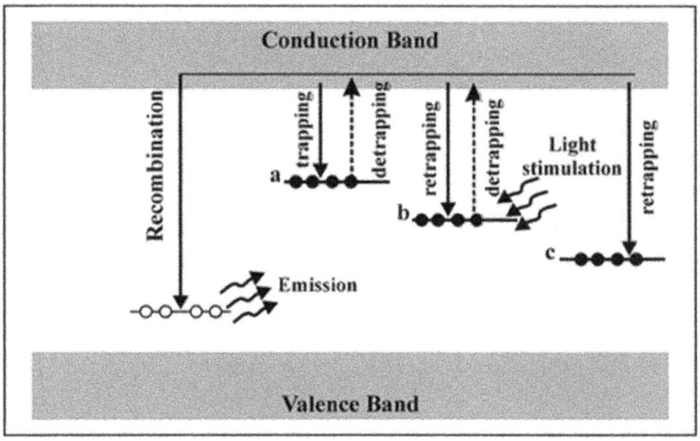

b

Figure 1.13. The electronic transitions among the dosimeter's energy bands during irradiation (a) and read-out (b) processes.

charges back to the conduction band, which results in electron–hole pair recombination and luminescence. The signal intensity of the stimulated luminescence is directly related to the trapped charge concentration and consequently the absorbed dose.

Apart from the conduction and valence bands there is also a hole trap that acts as a recombination center and three types of electron traps. The latter constitute shallow traps at level a, dosimetric traps at level b and deep traps at level c. The dosimetric traps (level b) are those that yield the luminescence signal used in OSL dosimetry. Figure 1.13(a) demonstrates the electronic transitions occurring upon irradiation, which create the radiation-induced electron–hole pairs and filling of traps. Figure 1.13(b) demonstrates the corresponding transitions during the read-out, where optical stimulation forces electrons to escape from traps and recombine with holes.

The trapped charges can be released to the conduction or valence band using optical stimulation. In case of optical stimulation, the escape probability *P* is given by the product of the photon flux (φ), in number of photons per unit time per unit area, with the photoionization cross-section (σ). σ describes the probability a photon at energy $h\nu$ to interact with a particular crystalline structure defect (Bøtter-Jensen *et al* 2003). Once the trapped charges escape, electron–hole recombination is possible resulting to OSL luminescence. The total OSL associated with a specific trapping level is proportional to the trapped charge concentration and, consequently to the absorbed irradiation dose.

A simple model for the OSL process assumes the absence of retrapping, that is, all charges escaping the traps recombine directly giving luminescence. Therefore, the luminescence intensity is proportional to the change of the trapped charge concentration over time as: $\frac{dn(t)}{dt} = -nP$, where $n(t)$ is the trapped charge concentration at time t.

The shape of the OSL intensity curve as a function of lighting stimulation time, can be obtained by solving the above equation using the escape probability *P* and a constant stimulation photon flux intensity φ as follows:

$$I_{\mathrm{OSL}}(t) \propto n_0 \sigma \phi e^{-\sigma \phi t} \tag{1.10}$$

The OSL intensity follows an exponential decay based on the change of the trapped charge concentration (McKeever 2001, Bøtter-Jensen *et al* 2003).

It should be noted that in real OSL dosimetry materials, a pure exponential decay is rarely observed. The OSL curve shape and signal intensity may be affected by various processes such as: trapping at competing traps, thermal stimulation of trapped charges, retrapping at the dosimetric trap, simultaneous stimulation from multiple trapping levels, recombination at multiple recombination centers and photo-transfer of charges from different energy levels.

1.3.2 Stimulation methods

Continuous wave OSL (CW-OSL) is the most commonly used method to stimulate OSL dosimeters. In CW-OSL a light source at an appropriate wavelength continuously illuminates the dosimeters while the OSL intensity signal is recorded. Pulsed OSL is another technique for OSL dosimetry that is used in low doses. In this technique the dosimeter is illuminated with bursts of short pulses of light from laser sources or LEDs, while intermediately monitoring the luminescence (Sanderson and Clark 1994). The pulse frequency is on the order of ~MHz and the OSL signal that is detected in-between the pulses is integrated over a large number of pulses. Appropriate selection of time triggering parameters enables the suppression of the signal associated with the scattered stimulation light, which improves the signal-to-noise ratio and the minimum detectable dose (McKeever *et al* 1996, Akselrod and McKeever 1999).

1.3.3 Basic experimental set-up

The instrumentation components required in OSL dosimetry are demonstrated schematically in figure 1.14. These include a light source, to be used for stimulation

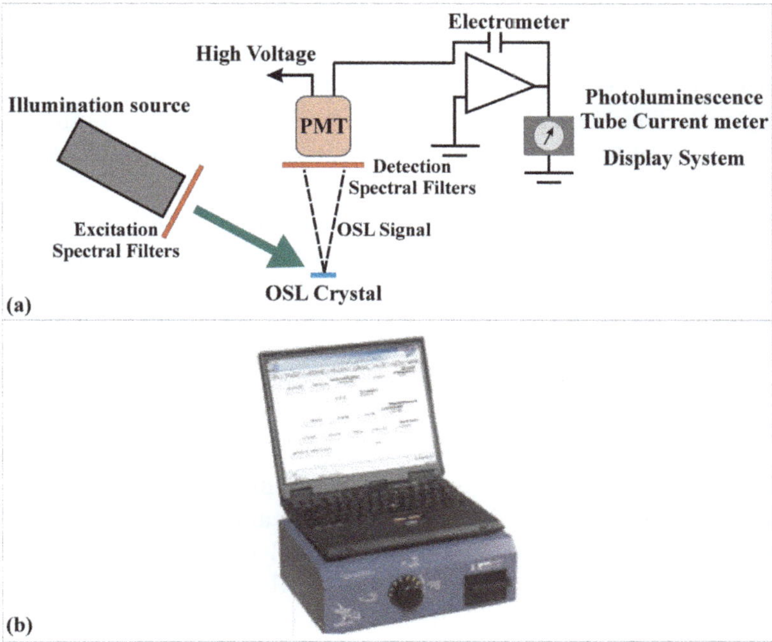

Figure 1.14. The configuration set-up of a typical OSLD reader (a), a commercially available OSLD reader device (b).

that is a laser or LED, spectral filters to select the wavelength of the light source and remove short-wavelength light, spectral detection filters to absorb stimulation light while transmitting the OSL signal from the dosimeter, a PMT for recording the light emitted by the dosimeter, and the required electronics. The intensity of the stimulation light is much higher than that of the OSL signal; so it is critical to use the correct spectral filtrations in front of the light source and the PMT to avoid detection of stimulation light. As an example, when using broadband LED sources such as green at 525 nm or blue at 470 nm, long-pass filters above 450 nm can be placed in front of the LED, and bandpass filters with a transmission between 290 and 370 nm can be placed in the form of the PMT (Bøtter-Jensen *et al* 2003). When laser light, such as the 532 nm from a frequency doubled Nd:YAG laser is used for stimulation, additional laser mirrors to block this laser line can be used to reduce the background signal. For optimal sensitivity, the selection of spectral filters has to take into account the emission spectrum of the light source and the emission spectrum of the OSL dosimeter.

1.3.4 The basic properties of Al_2O_3:C as an OSL dosimeter

Al_2O_3:C is produced in the presence of carbon, which creates a relatively high oxygen vacancy concentration and defects, known as F and F+ centers. The OSL sensitivity of the crystal is correlated with the concentration of F+ centers, which act as recombination centers (McKeever *et al* 1999, Akselrod and Akselrod 2002, Evans *et al* 1994). The F+ center concentration is of the order of 10^{15} cm^{-3}, while the

F center concentration is of the order of 10^{17} cm^{-3} (Akselrod *et al* 1990, McKeever *et al* 1999). The advantage of Al_2O_3:C is that it contains a dominant trapping level, which generates the OSL signal with the TL peak occurring at ~200 °C depending on the heating rate. This suggests that the OSL signal is stable at room temperature, while fading at room temperature is not considered appreciable at least for a period of up to 85 days (Bøtter-Jensen *et al* 1997).

The major emission spectral window of Al_2O_3:C is broadband and located at 420 nm. This is attributed to F center luminescence (Akselrod *et al* 1990, Markey *et al* 1995). During the OSL read-out process, electrons are optically released from trapping centers, recombining with the F+ centers. This recombination process generates excited F centers, which irradiate upon relaxation producing the recorded luminescence. Under specific conditions an UV emission band at 334 nm can also contribute the recorded signal (Yukihara and McKeever 2006b). For general applications, the OSL signal of Al_2O_3:C can be considered to be linear over a wide dose range in up to ~50 Gy if only the main luminescence band is detected. The UV emission band shows supralinearity starting at above 10 Gy, while sublinearity occurs for doses above 100 Gy (Yukihara and McKeever 2006b).

Apart from light sensitivity, a disadvantage of Al_2O_3:C is that its effective atomic number (11.28) differs to that of soft tissue (Bos 2001). This difference causes an over-response to low-energy X-rays compared to soft tissue (Akselrod *et al* 1990, Mobit *et al* 2006).

The Al_2O_3:C dosimeters that are commercially available are produced by Landauer Inc. and used in the Luxel™ and InLight™ systems. The Al_2O_3:C crystals are converted into powder and used to produce long plastic tapes with a total thickness of 0.3 mm, from which dosimeters of the desired shape and size can be obtained for use. The dosimeters in Landauer's Luxel™ badges are approximately 1.7 × 2.0 cm. InLight™ badges contain four round dosimeters 7.0 mm in diameter that can be read individually. The InLight™ Dot dosimeters contain a single round dosimeter 7.0 mm in diameter that can be placed in an adapter for read-out in the InLight™ or microStar™ reader. Long OSL strips are also being used to record the computed tomography dose index (CTDI) dose profiles in computed tomography (CT). The advantage of these dosimeters is the uniformity in sensitivity, since the Al_2O_3:C powder is a homogenized mixture of different crystal growth runs. Although the OSL signal can be erased by optical illumination (bleaching), sensitivity changes related to filling of deep traps in the crystal cannot be avoided.

Al_2O_3:C dosimeters are also available in single crystal forms. These dosimeters are typically 5 mm in diameter and 0.9 mm in thickness and are ideal for low dose measurements. Another advantage is that they can be annealed at high temperatures to empty deep electron and hole traps. A disadvantage is the large dosimeter to dosimeter variability, which is caused by the different defect concentrations present at each crystal. Thus, the OSL intensity, curve shape and dose response may vary from dosimeter to dosimeter, requiring individual crystal calibration for careful dosimetry. Smaller crystals have also been produced to be used as optical fiber probes by either cutting single crystals into smaller pieces of around

0.5 × 0.5 × 4 mm (Edmund *et al* 2006) or using other growth processes. These crystals would also be of interest for *in vivo* dosimetry in radiobiology experiments, a possibility that has yet to be explored.

1.3.5 Dose response

The dose response in OSL dosimetry depends on the experimental parameters employed and the type of dosimeter used. For Al_2O_3:C single crystals, the dose response may vary on the procedure followed for crystal manufacture (Yukihara *et al* 2004), on the spectral filters used in front of the PMT (Yukihara and McKeever 2006a, 2006b) and on the stimulation time window used to record the OSL signal (figure 1.15). The signal versus dose response may deviate from linearity when filling of deep electron and hole traps compete with the capture of free charges during either irradiation or read-out stage (Yukihara *et al* 2004, Yukihara *et al* 2003). The dose response can thus be affected by the irradiation history of the dosimeter owing to the filling of deep traps (Edmund *et al* 2006). A small supralinearity has been reported for doses above 2 Gy with radiotherapy photon beams (Jursinic 2007, Yukihara *et al* 2007). The sensitivity of the Al_2O_3:C dosimeters after multiple sequences of bleaching, irradiation and read-out has been shown to be within 0.6% of uncertainty for an accumulated dose of up to 2 Gy (Jursinic 2007). This result suggests that Al_2O_3:C dosimeters are reusable under the same experimental conditions.

The dependence of the Al_2O_3:C dose response has been investigated for photon beams at energies between 6 MV and 18 MV and for electron beams at energies between 6 MeV and 20 MeV. Measurement uncertainty has been reported to be less than 0.9% (Jursinic 2007, Aznar *et al* 2004, Yukihara *et al* 2007). The variation of

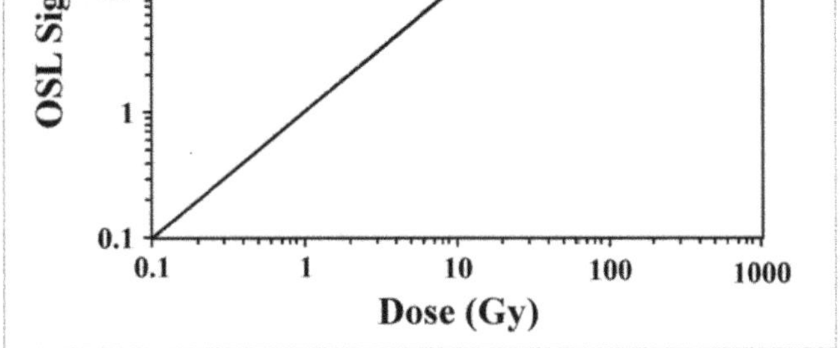

Figure 1.15. A typical dose response of the initial OSL signal intensity that is recorded during the first seconds of stimulation and the total OSL signal intensity that is detected from over 500 s of stimulation. Adapted from Yukihara *et al* (2004).

the beam energy with depth in water for both photons and electrons did not affect precision in the determination of the depth dose curves. A comparison the depth dose curves generated with Al_2O_3:C and an ionization chamber showed a difference of less than ±1%.

OSL dosimeters show no dependence on the angle of irradiation (non-directional dependence), dose rate and field size (Jursinic 2007, Yukihara *et al* 2007, Viamonte *et al* 2008).

1.3.6 Transient signals

OSL dosimeters emit a transient signal with a half-life of about 1 min after irradiation This signal may be related to shallow traps. Moreover, a higher and time-dependent signal can be observed due to phosphorescence caused by shallow traps immediately after irradiation or due to the fact that the shallow traps are filled following irradiation (Jursinic 2007, Pagonis *et al* 2006) As a result they do not act as effective competitors for the capture of charges stimulated from the dosimetric trap. A several minutes interval delay between irradiation and read-out should thus be kept in OSL dosimetry to avoid interference of the above unwanted signals.

1.3.7 OSL dosimetry in diagnostic radiology

The use of OSL dosimeters in diagnostic radiology is similar to TLD dosimetry. The main limitation of Al_2O_3:C when used for *in vivo* dosimetry is its over-response to low-energy X-rays with respect to tissue, caused by its effective atomic number (see section 3.1.8). However, this limitation may be avoided when dosimeter calibration is performed at an energy similar to that of measurement.

Few studies have been performed on the application of OSL dosimeters in CT. Authors have investigated the dose linearity of a Al_2O_3:C and a KBr:Eu optical fiber based real time dosimeter with tube current (mA) at various tube voltages (kVps) in air, at the center of a CTDI body phantom and at different radiosensitive organ sites inside an anthropomorphic phantom. The dose response was found to be linear with mA at a standard deviation of measurement of less than 6%. The profile of the X-ray beam at the center of the CTDI body phantom was measured using a series of OSL dosimeters at different positions along the *z*-axis of the phantom. Using a KBr:Eu OSL optical fiber based system, the results were obtained in real time with a precision better than 3%. However, both Al_2O_3:C and KBr:Eu exhibit energy dependence, which varied by up to 30% in the range from 80 kVp and 140 kVp. Landauer Inc. offers measurement of the dose profile in CT scanners using 15 cm long OSL strips. The strips are inserted along the five holes of the CTDI head or body phantom. After exposure, the OSL strip may be read out to record the dose profile at a resolution of as low as 0.1 mm. The CT dose index ($CTDI_{100}$) may then be calculated. Beam profiles obtained from helical scans using these OSL strips have been used to verify measurement fields along the axes of the CTDI body phantom and to compare the dose distribution with typical measurements obtained from an ionization chamber. Aznar *et al* (2005a) have used an Al_2O_3:C optical fiber based OSL system to determine the entrance and exit doses in mammography. The dose response of the dosimeter was linear over the range investigated (4.5–30 mGy) and the reproducibility of the measurements at an air kerma of 4.5 mGy was 3%. The dose response was, however, increased by 18% with an increase in the kVp from 23 to 35 due to the photon energy dependence of Al_2O_3:C. This energy dependence was simulated using Monte Carlo methods by Aznar *et al* (2005b). The authors demonstrated that the OSL probes did impair the evaluation of the mammography images due to their distinct shape and small size.

1.3.8 A comparison of the TL versus OSL dosimetry

OSL dosimetry has many advantages compared to TLD dosimetry. OSL requires light to retrieve the dose record. This process is significantly faster and more efficient compared to the heating required in TLDs. Light illumination can be controlled much more precisely than heat. Thus, the dose measurements from OSL dosimeters are considered to be more accurate than TLDs. OSL dosimeters are sensitive to very low and very high doses. Typical OSL dosimeters can be used to record doses from 10 μGy to several tens of Gy. Most of the energy stored in the TLD material is depleted upon heating, leaving no possibility to re-read the material. This is not the case with OSL dosimeters since they release only a fraction of the absorbed energy when optically stimulated. Thus, an OSL dosimeter can be read out a number of times, which reduces the uncertainty of the results. OSL crystals can be manufactured in a variety of forms and shapes such as powder and fibers. They are also highly stable in terms of variations in temperature and humidity. Table 1.5 demonstrates the comparative characteristics in TLD versus OSL dosimetry.

Table 1.5. Comparative characteristics in TLD versus OSL dosimetry.

TLDs	OSLDs
High sensitivity, subject to thermal quenching and irradiation background	High sensitivity, absence of thermal quenching and irradiation background
Single reading out	Multiple reading out to reduce measurement uncertainty
Read-out time ~minutes	Read-out time ~1–2 s
Easy handling; no light sensitivity	Highly light sensitive but can be avoided by appropriate precautions in use and storage
Dose range: 1 μGy to 100 Gy	Dose range: 10 μGy–10 Gy
Measurement of dynamic real time exposure is difficult	Measurement of dynamic real time exposure is possible
Post-irradiation fading: <2% in 6 months	<5% in a year

1.4 Computational phantoms simulating pregnancy and children

1.4.1 Introduction to computational phantoms

The use of physical phantoms in dosimetry has proven to be laborious due to the experimental radiation procedures as well as the time needed to process the dose results. The advent of the first generation of computers and Monte Carlo simulation algorithms made the calculation of dose deposition to human body possible using computational methods. Computational human phantoms are mathematical descriptions or models that represent organs or tissues of the body or the whole body itself. Coupled with information about tissue elemental composition and density, researchers are able to trail radiation interactions and energy deposition inside the body. Although additional information must be supplied for the radiation source and detectors, Monte Carlo statistical methods and computational phantoms are extremely advantageous in comparison with the physical phantom approach, especially in internal dosimetry. Detailed computational representations of the human anatomy became popular as soon as 3D imaging techniques were developed. These techniques enabled the production of cross-sectional views and 3D-dimensional images of internal organs and structures of the body, without the superposition of images of structures present in 2D-dimensional medical radiography. During the 40 last years, the development and application of computational models for radiation dosimetry has grown into a unique specialized field of research in radiation protection, dosimetry and radiotherapy. In fact, today, computational models are the key elements in designing, testing and evaluating both physical phantoms, detectors and acquisition technologies.

1.4.2 Evolution and basic components of computational phantoms

Computational models for human anatomy have existed from more than 50 years. Since the 1960s over 100 computational phantoms have been reported in the literature to complement over 25 physical phantoms involved in ionizing and non-ionizing applications (Xu and Eckerman 2010). In 1996, the first workshop on voxelized computational phantoms was organized at the National Board of Radiological Protection (NRPB, now HPA) in the UK (Dimbylow 1996). In 2000, a similar workshop was hosted at Oak Ridge National Laboratory (ORNL) in the United States (Eckerman 2000). The interest in computational phantoms was so prevalent in the 2000s that many researchers decided to found the Consortium of Computational Human Phantoms, an international initiative to promote research in computational modeling of the human body for medical and occupational organ doses related to exposure to radiation. The organ and body surfaces of computational phantoms have been characterized with a variety of solid geometry techniques: quadric equations, voxels and advanced primitives such as B-splines or non-uniform rational B-splines (NURBS) or meshes. The advance of these techniques reflects the progress in computer and medical imaging technologies during the last 50 years of phantoms' evolution. In general, 3D computational objects are described by two major components: geometry and material constituents that represent the exterior features/

boundaries of the human body, detailed internal tissue structure, tissue density and chemical composition, allowing for a transport code to simulate interactions and energy deposition.

Constructive Solid Geometry (CSG) allows the modeler to create solid objects to combine very simple shapes such as cuboids, cylinders, prisms, pyramids, spheres, cones and ellipsoids—surfaces that are easily described by quadric equations. Representations are easy to adopt and can yield good results when the objects are relatively simple in shape. Topological information can include relations (intersections) and orientation between surfaces. Surface equations are computationally efficient and can be used in Monte Carlo codes. Unlike the CSG representation, boundary representation (BREP) is much more flexible because a richer set of operation tools are available. These features allow BREP-based models to include very complex anatomical features. Topological information provides the relationships among vertices and edges of the surfaces. In BREP, the exterior of an object can be defined as non-uniform b-spline functions, so the model can have very smooth surfaces. Voxel phantoms are based on the discretization of an object into tiny cubes, called voxels. The source of voxel phantoms are the 3D scanning technologies: CT, MRI (magnetic resonance imaging), PET (positron emission tomography), as well as cross-sectional high-resolution photography of cadavers through which detailed human anatomy may be acquired. Images obtained with these techniques represent a matrix of voxels, where each voxel belongs to a specific organ or tissue. By using segmentation algorithms, voxels representing specific tissues and organs are placed within the 3D matrix, comprising the voxel phantom. To simulate the radiological properties of the human body, a meticulous composition modeling of tissues is required. Tissue equivalence can be expressed by parameters such as physical density and effective atomic number (Z_{eff}); however, electron density and mass absorption coefficient are widely acknowledged to accurately represent the mechanisms of energy deposition for any radiation interaction pattern.

1.4.3 Solid geometry stylized phantoms

The first solid geometry age-dependent stylized phantoms with body surface and organs expressed by mathematical equations, to simulate heterogeneous human tissue, were based on the initial MIRD phantoms developed at ORNL in the 1970s and 1980s (Snyder *et al* 1978, Cristy 1980). The original MIRD phantom was assumed to be tissue-distributed homogeneously throughout. Regions of little dosimetric importance were not included such as fingers, ears and nose. In addition, no attempt was made to model the lungs or skeleton or to define the locations of specific organs or tissues in the phantom. The representation of internal organs in this mathematical phantom was rough and unsophisticated, as simple equations can express only the general attributes of the position and geometry of each organ. The original model was intended to represent a healthy 'average' adult male, and female, a newborn, and individuals of ages 1, 5, 10, and 15 developed from anthropological data (legs, trunk, and head) and from age-specific organ masses published in ICRP Publication 23. The Reference Man was a 70 kg, 174 cm height, 25-year-old Caucasian. In 1978, Snyder *et al* published an improved version of their heterogeneous phantom that contained more than 20 organs and detailed anatomical features. These phantoms underwent numerous modifications to improve radiation dosimetry. At the same time, significant efforts were undertaken at ORNL during the mid-1970s to develop individual paediatric phantoms based upon a careful review of the existing literature for each age. Three 'individual phantoms' were designed by Hwang *et al*. This set consisted of the newborn, the one-year and 5-year-old models. A separate effort was undertaken at ORNL by Jones *et al* for a 15-year-old model, so that four designs were complete (Jones *et al* 1976). Each phantom composed of three tissue types with different densities: bone, soft tissue and lung. During the 1990s, Stabin adapted the ORNL adult female phantom to develop three pregnant models at the end of each trimester of pregnancy. The 3 months uterus was represented by a right circular cone with a hemispherical cap with a 33° inclination. In the 6 months pregnant phantom, the uterus is modeled as a cylinder capped at both ends by hemispheres. The long axis of the cylinder is tilted upwards at an angle of 40° from the horizontal. In the 9 months pregnant phantom, the uterus was modeled likewise, to have extended length, and size. No attempt was made to model skeletal tissue in the fetus at that time. On the other hand, the National Research Center for Environment and Health in Germany (GSF) center used the anatomical descriptions of the hermaphrodite MIRD-5 phantoms to develop a pair of gender-specific adult phantoms known as the ADAM and EVA including differently designed, sex-specific organ such as testes, ovaries and uterus (Kramer *et al* 1982). Also, a different chin was introduced by removing a section of the neck to create a more realistic external irradiation geometry for the thyroid. There were several anatomical differences between GSF and ORNL phantoms, such as the breast surface equation and sizes. Referring to Korean reference man anthropomorphic data, phantoms of a Korean Adult Male and Female were constructed based on modified mathematical surface equations representations of the ORNL adult computational phantom (Chen 2004). Later on, a group at the Nagoya Institute of Technology developed a new 9 month infant stylized phantom for non-ionizing radiation dosimetry (Hirata *et al* 2008). With the availability of general-purpose Monte Carlo codes and affordable computers in the 1980s and 1990s, these series of phantoms were adopted and used for a wide variety of internal and external dosimetry applications. For nearly 40 years since the first anthropomorphic phantom was reported, all these anatomically simplified representations have been utilized as the standard radiation dosimetry 'Reference Man' approach, which is based on 'population-average' 50th percentile anatomical parameters (ICRP 1975). Although stylized phantoms made it possible to carry out Monte Carlo computations during times when computers were much less powerful, human anatomy is too complex to be credibly represented with a limited set of equations. Many anatomical details forfeited in the above models sometimes led to erroneous results (table 1.6).

Table 1.6. Popular stylized computational phantoms including information on the phantom names, anatomical features modeled and literature references.

Phantom name(s)	Anatomical features/human subjects	Reference/developer (s)
ORNL MIRD-5	Hermaphrodite adult, ICRP 23	Snyder *et al* (1969, 1978)
Cristy Eckerman family phantoms	Caucasian adult, newborn, 1 year, 5 year, 10 year, 15 year old	Cristy (1980, 1987)
–	Pregnant phantom at 3, 6, 9 months of gestation	Stabin *et al* (1995)
GSF ADAM & EVA	Caucasian adult male and female	Kramer *et al* (1982)
K-MIRD	Korean adult male	Park *et al* (2006)

1.4.4 Voxelized phantoms

The evolution of tomographic imaging technologies such as CT and MR imaging, enabled researchers to visualize the complex structure of the human body with remarkable resolution, originating the era of tomographic phantoms. In terms of the development process, tomographic phantoms are fundamentally diverse to stylized ones. To construct a whole body phantom, image slices should ideally cover the whole human body, which is a procedure not normally performed in routine medical examinations. As a result, the data size of a whole body model, especially when high-resolution scans are involved, can be potentially too large to handle. In addition, a large amount of internal tissues need to be identified and segmented. Moreover, considerable Monte Carlo simulation capabilities are often required for diverse radiation types. A tomographic image data set is composed of many slices, each displaying a two-dimensional pixel map of the anatomy. Unlike stylized phantoms, which are often based on quadratic surface equations, a tomographic phantom contains a huge number of voxels (cuboids) grouped to represent various anatomical structures. The creation of a tomographic phantom involves four general steps: (1) acquisition of tomographic images covering the entire volume of the body; (2) identification and segmentation of organs or structures of interest from the original image slice; (3) specification of the density (e.g., soft tissue, hard bone, air, etc) and chemical composition of segmented tissues; and (4) registration of the segmented data into a 3D volume that can be used for 3D visualization as well as implementation in a specific Monte Carlo transport code. Zubal *et al* (1994) reported a head-torso model developed from CT images and divided the original data to two sets by attaching arms and legs in two different positions to the original torso phantom. Kramer *et al* (2006) from Brazil developed an adult male and later a female phantom named FAX, both adjusted in accordance with ICRP-89 reference body heights and organ masses, and later revised their skeletons to improve their compatibility with the latest ICRP-103 report recommendations. NRPB reported the development of an adult male phantom from MRI sequences, with a body height

Table 1.7. Popular voxelized computational phantoms including information on the phantom name(s), anatomical features modeled, and literature references.

Phantom name(s)	Anatomical features/human subjects	Reference/developer (s)
Zubal phantom	Caucasian adult male head and torso	Zubal *et al* (1994)
FAX, MAX	Caucasian adult male and female	Kramer *et al* (2004)
NRPB NAOMI & NORMAN	Caucasian adult female and male, pregnant at various gestation stages	Dimbylow *et al* (1997, 2005, 2006)
UF series A, B, 0 year, 2 months	Paediatric phantom series/ICRP-89/ 9 months – 14 year	Nipper (2002), Lee (2005, 2006)
GSF series REGINA and REX	Adult female and male/ICRP-89, Caucasian 43 year and 38 year	ICRP publication 110 (2009)

like the ICRP Reference Man, to determine the specific energy absorption rate from exposures to non-ionizing radiation. Later, NRPB developed an adult female version, named NAOMI, rescaled to the dimensions of the ICRP Reference Woman. Later, they merged the NAOMI phantom with the stylized fetal phantoms developed by Chen to create a series of hybrid phantoms for pregnant women. In addition, additional paediatric voxel phantoms were developed from the UFL representing children with ages ranging from newborn to 15 years old (Lee *et al* 2005). This approach was later extended to two groups of phantoms; one group similar to ICRP Publication 89 reference values and another group of scaled down phantoms of 1 year age intervals, from newborn to 15 years old, to provide a library of phantoms for age-specific organ dose assessment. Shi and Xu from Rensselaer Polytechnic Institute (RPI) also published a pregnant woman phantom, developed from a unique body CT of a pregnant patient at the last trimester of gestation (Shi and Xu 2004). Also, GSF used CT images of healthy volunteers to develop what eventually became a family of 12 voxel phantoms. Among these phantoms, two phantoms underwent revision to produce the REGINA and REX phantoms, which were released to the public as the ICRP Reference Computational Phantoms (Zankl *et al* 2005). The organ masses of both models were adjusted to the ICRP data on the adult Reference Male and Reference Female, without compromising their anatomical realism. Subsequently, the ICRP released its Publication 110, a major step forward into standardization, which describes the development and intended use of these so-called ICRP computational phantoms. These phantoms are created from real medical image data, nonetheless they are consistent with the data given in ICRP-89 on the reference anatomical and physiological parameters for both male and female subjects (table 1.7).

1.4.5 Boundary representation phantoms

Because voxelized phantoms are usually built from images of specific individuals, there is a deficiency of anatomical alterability associated with organ size, shape and location, which are important qualities in dosimetry. An approach to resolving this limitation is the use of NURBS and/or polygon mesh surfaces. This approach utilizes both the anatomic realism of voxel phantoms and the mathematical flexibility of stylized phantoms. The first step is segmentation of organs and tissues from CT or MRI patient images, similar to the procedure performed for the creation of voxel phantoms. Once the voxelized anatomy is obtained, it is transformed into a mesh model, where both organs and body contours are represented by an array of polygon surfaces. The next step is to transform these meshes to NURBS surfaces. The data set obtained phantom is called a hybrid array, since it is composed of NURBS and polygon mesh surfaces. The fourth step is related to voxelization of the phantoms. Voxelizing a hybrid phantom reintroduces most of the limitations of the voxel phantoms. This step is required for making the data set compatible with MC radiation transport codes, and it is usually performed by filling the NURBS/polygon mesh structures with cubic voxels of user-defined size. Several groups have reported polygon mesh phantoms and techniques for BREP phantoms. RPI reported the development of a series of phantoms representing a pregnant woman and her fetus at the end of each trimester of gestation (Xu *et al* 2007). These phantoms, referred to as the RPI Pregnant Females, described by polygonal meshes, were derived from binding anatomical information from a nonpregnant female, a 7 month pregnant woman CT data set, and a model of the fetus. The RPI group also reported the development of a pair of adult male and female phantoms that was adjusted to match the ICRP-89 reference values for numerous organ and bone structures (Xu *et al* 2007, 2008). The UF group created a BREP phantom series, so-called UFH-NURBS phantoms using patient CT image data segmentation adaptation to polygon meshes converted to NURBS format. These meshes were then converted to the NURBS format using commercial software. In 2008, a group led by Stabin, published a family of adult and paediatric phantoms by adapting the NURBS-based NCAT adult male and female phantoms (Stabin *et al* 2008). All preceding phantoms were ICRP-89 reference body- and organ-adjusted (table 1.8). All these approaches have the following advantages: (1) the NURBS-based phantoms can be created much more quickly than voxel phantoms and manually segmented individual patient image data sets; (2) the phantoms have a higher level of internal regularity; and (3) the phantoms are complete from head to toe.

Table 1.8. Popular BREP computational phantoms including information on the phantom name(s), anatomical features modeled, and literature references.

Phantom name(s)	Anatomical features/human subjects	Reference/developer(s)
RPI polygon mesh phantom series	Pregnant 3, 6, 9 months, ICRP-89	Xu *et al* (2007)
UF NURBS phantom series	Caucasian 6 day old female, 14 year male, 14 year female	Lee *et al* (2005, 2006)

1.5 Monte Carlo simulations and computational phantoms

1.5.1 Deterministic and probabilistic calculation methods

Besides deterministic solution of the radiation transport problem, which is based on solutions to Boltzmann's equation, the transport problem can be solved through a probabilistic formulation.

Monte Carlo methods, as utilized in radiation transport applications in radiotherapy and dosimetry, provide a numerical solution to the Boltzmann transport equation that employs the fundamental microscopic physical laws of electron–atom and photon–atom interactions. The fluences of individual particle tracks are reproduced within the physical laws of scattering and absorption cross-sections. Macroscopic features of the radiation fields, for example, the average track-length per incident photon, are computed as an average over many individual particle simulations. The Central Limit Theorem indicates that if a true average value of x exists and the distribution of x has a true finite variance σx^2, the Monte Carlo estimator for a mean value can be made arbitrarily close to real x, by increasing the number of particle histories simulated and also predicts that (a) the distribution of x is Gaussian characterized and (b) the variance σx^2 will, proven by the Strong Law of Large Numbers, approach zero if the number of particles approaches infinity.

1.5.2 Introduction to Monte Carlo methods

Commonly, the first reference to Monte Carlo methods is that of Comte de Buffon (Buffon 1777), who proposed a method to evaluate the probability of tossing needles onto a ruled piece of paper.

Many years later, Laplace in 1886 proposed a procedure that could be employed to approximate the value of π, by repeatedly throwing a needle onto a lined sheet of paper and counting the number of intersected lines. Several other historical uses of Monte Carlo methods predating computers are cited by Kalos and Whitlock (1986). The modern Monte Carlo age was brought by pioneers in the development of the Monte Carlo technique, von Neumann and Ulam, who achieved its realization in digital computers during the initial development of nuclear fission weapons. The Monte Carlo term was given by Von Neumann because of the similarity of statistical simulation to games of chance, and because the principality of Monaco was a center for gambling and betting quests. The systematic development of the method dates from the last 50 years with various studies and reports, describing the principles of the Monte Carlo method and its applications in medical radiation physics. There has been an enormous increase and interest in the use of Monte Carlo techniques in all aspects of medical physics, including diagnostic radiology, radiotherapy, radiation dosimetry, nuclear medicine and radioprotection.

1.5.3 Monte Carlo radiation transport codes

In general, Monte Carlo simulation of particle transport requires a great deal of information regarding the interaction properties of the particle and the media through which it travels. A particle history is begun by creating the particle with position and energy information according to a specified source distribution. The particle travels a certain distance before undergoing an interaction. The type of interaction and resulting particles are determined by the interaction cross-sections at that point. Any secondary particles created are also accounted for and the particle history ends when all particles have either deposited their energy within the medium or have left the region of interest inside the medium. The main steps involved in ionizing radiation transport through a certain medium are given in figure 1.15.

The main components of a Monte Carlo code may contain: (i) the probability density function: the idea of how the behavior of a system can be described by a curve called the probability density function, and how the properties of that curve can help to simulate its behavior; (ii) the random number generator: a source of random numbers uniformly distributed on the unit interval; (iii) a prescription rule for sampling the specified parameter; (iv) score keeping or tallying: the outcomes must be accumulated into overall scores for the quantities of interest; (v) an estimate of the statistical error variance as a function of the number of trials; (vi) variance reduction techniques: methods for reducing the variance in the estimated solution to reduce the computational time for Monte Carlo simulation (figure 1.16).

A series of Monte Carlo codes have been developed. The most widely used general-purpose Monte Carlo packages are described below in this section. The Monte Carlo N-particle (MCNP) code is developed and maintained at Los Alamos National Laboratory and is distributed via the Radiation Safety Information Computational Center. MCNP has been used successfully to model spectra from medical linear accelerators, estimate the dose distributions around high dose rate brachytherapy sources, and assess the dosimetric properties of new radioactive sources. MCNP has also been used to predict fast neutron activation of shielding and biological materials. In diagnostic applications, MCNP has been used to model X-ray CT and positron emission tomography scanners, to compute the dose delivered from CT procedures, and to determine detector characteristics of both diagnostic and nuclear medicine devices. MCNP has also been used to determine particle fluxes around radiotherapy treatment devices and to perform shielding calculations in radiotherapy treatment rooms and diagnostic X-ray facilities.

The GEANT code was originally indented for high energy physics experiments but has found applications in the areas of medical sciences and radiation protection. Geant4 can manage transport of electrons, positrons, γ-rays, X-rays, optical photons as well as hadrons and ions' interactions used in particle cancer treatments. GEANT4 is developed and maintained at CERN, being a result of a large collaboration of the scientific community.

FLUKA is a general-purpose MC radiation transport code that tracks nearly all particles (photons, leptons and hadrons) over an extended energy range. The present multi-particle, extended energy code started in 1989 as an effort mainly of the National Institute for Nuclear Physics (INFN, Italy) and has been in continuous development, in the frame of collaboration between INFN and CER. The main applications in medical radiation physics remain in hadron therapy, dosimetry and radiation protection.

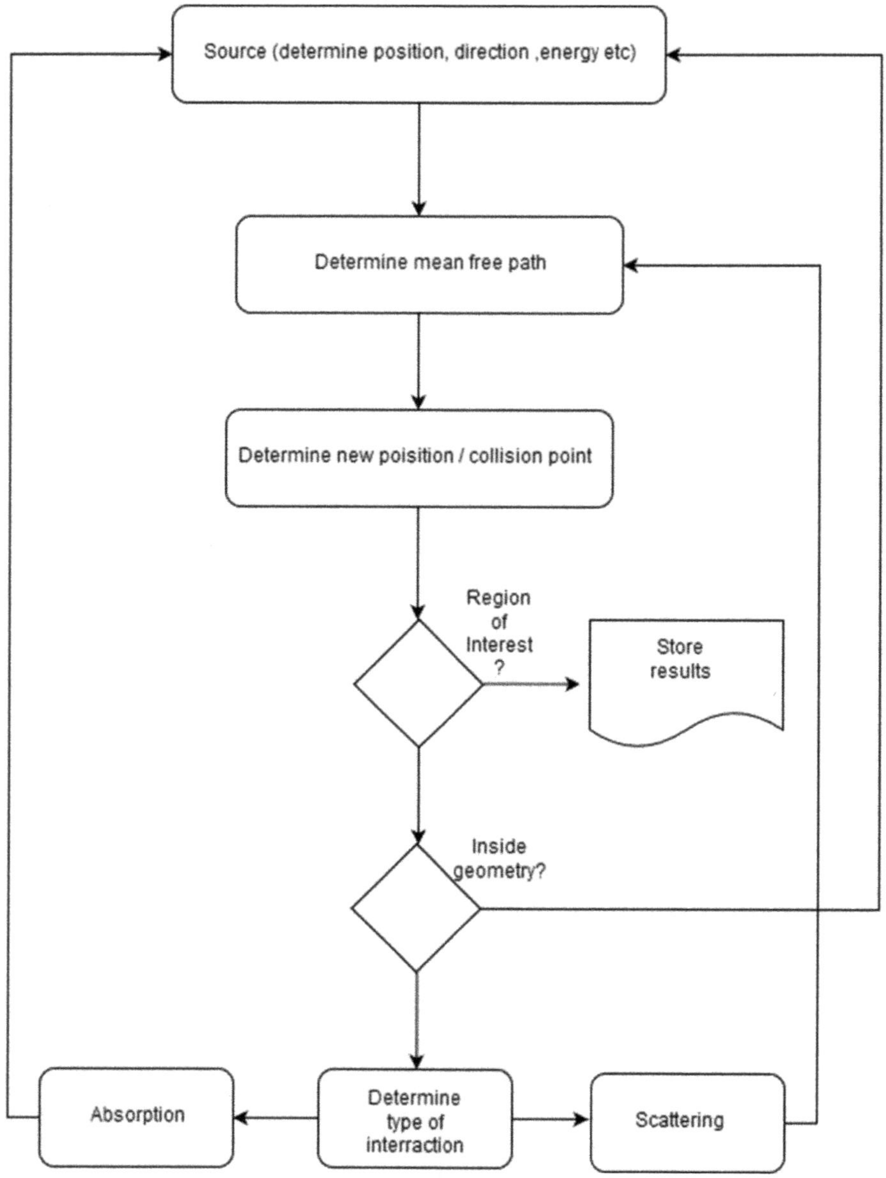

Figure 1.16. Basic flowchart of a Monte Carlo radiation transport method. MCNP and other radiation transport codes.

PENELOPE is a code system that performs MC simulations of coupled electron–photon transport from a few hundred eV to about 1 GeV. Photon transport is simulated by means of the standard, detailed simulation scheme. Electron and positron histories are generated based on a mixed procedure, which combines detailed simulation of hard events with condensed simulation of soft interactions. A geometry package called PENGEOM permits the generation of random electron–

photon showers in material systems consisting of homogeneous bodies limited by quadric surfaces, that is, planes, spheres, cylinders, etc. PENELOPE was developed at the University of Barcelona and is distributed via the Nuclear Energy Agency.

1.5.4 Commercially available computational dosimetry software

Apart from the dedicated Monte Carlo codes mentioned in the previous section, end-users can use out-of-the-box Monte Carlo software for simulating particle and energy transport for radiation dosimetry in simple or complicated medical X-ray examinations. Some of this software is based on uncomplicated stylized phantoms, other software is based on complex voxelized models for patient-specific dosimetry or based on pre-calculated dose coefficients for evaluating energy deposition in patient organs.

PCXMC is a Monte Carlo program for calculating patients' organ doses and effective doses in X-ray examinations. The program calculates the effective dose with both the present tissue weighting factors of ICRP-103 (2007) and the old tissue weighting factors of ICRP 60 publication. The anatomical data are based on the mathematical hermaphrodite phantom models of Cristy and Eckerman, which describe patients of six different ages: newborn to 15 years of age and adult. Some changes were made to these phantoms in order to make them more realistic for external irradiation conditions and to enable the calculation of the effective dose utilizing the ICRP-103 tissue weighting factors. The phantom sizes are adjustable to mimic patients of an arbitrary weight and height. PCXMC allows a free adjustment of the X-ray beam projection and other examination conditions of projection radiography and fluoroscopy. The Monte Carlo calculation of photon transport is based on stochastic mathematical simulation of interactions between photons and matter. Photons are emitted from a point source into the solid angle specified by the focal distance and the X-ray field dimensions and followed while they randomly interact with the phantom according to the probability distributions of the physical processes that they may undergo: photo-electric absorption, coherent (Rayleigh) scattering or incoherent (Compton) scattering. At each interaction point the energy deposition to the organ at that position is calculated and stored for dose calculation. Other interactions are not considered in PCXMC because the maximum photon energy is limited to 150 keV. This chain of interactions forms a so-called history of an individual photon. A large number of independent photon histories is generated and estimates of the mean values of energy depositions in the various organs of the phantom are used for calculating the doses in these organs.

ImpactMC is a software package providing fast calculation of 3D dose distributions for radiography and CT scans using Monte Carlo algorithms. The generated 3D dose distributions are useful to estimate organ doses and to calculate the effective dose of individual scans. Based on the parameters of a scan (X-ray spectrum and geometry) and its reconstructed volumetric data, a Monte Carlo simulation is performed in order to calculate the dose deposited in each single volume element (voxel) considering all relevant photon interaction processes. ImpactMC's output comprises parametric dose images in which every voxel carries the normalized dose (to CTDI) of the initial image grid. The software can be applied to a wide range of applications from radiography to clinical CT. Each estimation of the dose distribution in a CT volume starts with setting up an ImpactMC project. An ImpactMC project is a collection of parameters and links to data files that contain information about the simulation to be performed. After creating a new project, the user needs to include the input CT volume, scanner and scan protocol characteristics, including collimation, shaped filter, X-ray spectrum and tube current modulation, if used. In addition, information is needed for Monte Carlo simulation characteristics, including conversion from CT values to material types and density values and user-defined material types. The user can enter the necessary input parameters using a straightforward GUI.

CODE (embryodose.med.uoc.gr) is a free web-based software tool developed in the Department of Medical Physics, UOC, dedicated to estimate conceptus radiation dose in case of (a) pregnant patients subjected to radiological examinations and (b) pregnant employees exposed during fluroscopically guided interventional procedures. Also, conceptus risk for radiation-induced cancer following various examinations performed on the mother can be calculated through a facile GUI, oriented not only for physicists but for clinicians as well.

The CODE radiography module provides estimates for the embryo absorbed dose and associated risk for childhood cancer from conventional radiographic projections performed on the pregnant patient. MCNP transport code and three mathematical phantoms representing the pregnant individual at first, second and third trimester of gestation were employed to produce normalized embryo dose (ED) data from common radiographic projections performed on the expectant mother. Normalized to skin surface dose embryo dose data (NED) for each radiographic projection have been produced for several combinations of tube potential and total X-ray tube filtration. In addition, NED data for the first trimester of gestation were generated for various embryo depths below the anterior abdominal surface. The dose to the ED from a series of radiographic projections is calculated from the equation:

$$\mathrm{ED} = \sum_i^j \left\{ \left[\left(\frac{\mathrm{FDD}_O}{\mathrm{FSD}_X} \right) \cdot \left(\frac{\mathrm{kV}_X}{\mathrm{kV}_O} \right) \right]^2 \cdot I_O \cdot \mathrm{mA} \right\}_i \cdot \mathrm{NED}_i \tag{1.11}$$

where NED_i is the normalized ED for the i radiographic exposure performed and exposure parameters of the examination (tube voltage kV_X, total filtration and focus to skin distance FSD_X). I_O is the tube output measured at FDD_O for a tube voltage of kV_O.

For fluoroscopy, normalized ED data were produced from common fluoroscopic projections involved in FG procedures performed on the expectant mother. The embryo dose was normalized to the dose area product (DAP). Based on the derived NED data, fitting equations were produced for the estimation of ED for any tube potential, filtration and embryo depth.

The dose to the ED from a series of fluoroscopic projections is calculated from the equation:

$$\mathrm{ED} = \sum_{i}^{n}(\mathrm{DAP}_i \cdot \mathrm{NED}_i) \tag{1.12}$$

where DAP_i is the DAP recorded for the projection i performed at a specific tube voltage, total filtration and focus to skin distance, and NED_i is the normalized to DAP ED for the i fluoroscopic projection and the same exposure parameters. For both radiographic and fluoroscopic procedures, fitting equations were produced for the estimation of ED for any tube potential, filtration and embryo depth. Also, the radiogenic risk for childhood cancer associated with in-utero exposure can be estimated using – as recommended by the International Commission on Radiological Protection – report 90.

The CODE CT module provides estimates for the embryo absorbed dose and associated risk for childhood cancer from CT examinations of the trunk performed on the pregnant patient. Normalized to free-in-air CTDI ED coefficients ($f(z)$) for single sequential scans at different positions along the z-axis to cover the whole trunk of the pregnant individual were produced through Monte Carlo simulation experiments. The MCNP transport code and four mathematical phantoms representing the average pregnant individual at first post-conception weeks (0–7 weeks), first trimester (8–12 weeks), second trimester (13–25 weeks) and third trimester (26–40 weeks) of gestation were employed.

The ED from a specific CT exposure performed on a pregnant patient is estimated using the formula

$$\mathrm{ED} = \mathrm{CTDI}_F \cdot \mathrm{NCD}_{p_0} \cdot f_{p,d} \cdot f_{\mathrm{scanner}} \tag{1.13}$$

where

a. CTDI_F is the free-in-air CTDI (mGy) of the user's CT scanner for the tube voltage, tube load and beam collimation employed.
b. NED_{p0} is the cumulative normalized ED for the specific boundaries of the scanned body region modified for the beam collimation and pitch used for the examination, where p_0 is the abdominal circumference of the average pregnant individual at the gestational stage of the examined pregnant patient.
c. $f_{p,d}$ is the correction factor for embryo depth d and abdominal circumference p.
d. f_{scanner} is the correction factor for the specific scanner used for the examination.

More information about CODE can be found in chapter 5.

1.5.5 Computing conceptus and paediatric doses using Monte Carlo radiation transport codes

Using a general-purpose MC transport may prove to be more laborious compared to off-the-shelf dosimetry software with or without MC integration. The user must

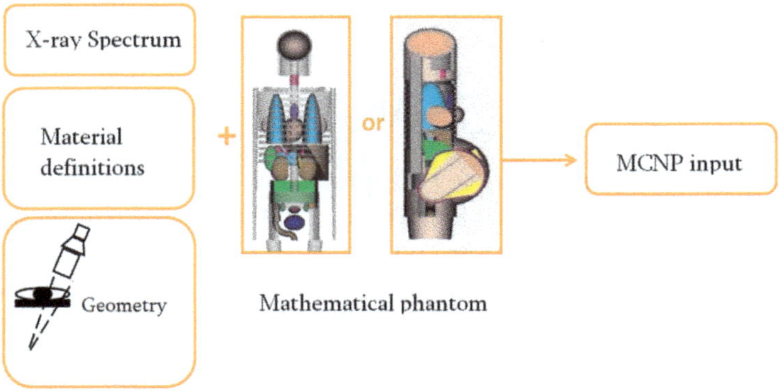

Figure 1.17. The MCNP input file is a formatted representation of the X-ray spectrum, the irradiation geometry and the mathematical phantom with materials' definition.

supply an input file that is subsequently read by MCNP. This input file contains information about the dosimetric problems to be solved/simulated, such as the geometry specification, the description of materials and selection of cross-section evaluations, the locations and characteristics of the photon source(s), the type of answers or tallies desired and any variance reductions techniques to improve the efficiency of the calculations (figure 1.17).

In MNCP coding, each finite medium that is filled by a determined material is called a 'cell'. Any cell can be defined with surrounding surfaces. All space must be defined as having non-zero importance or zero importance and at least one cell must be defined and be 'syntactically' correct for the dosimetric problem to be valid for simulation. The cells are defined by the intersections, unions and complements of the regions bounded by the surfaces. Surfaces can obviously be defined as two- or three-dimensional equations to specify the areas that enclose the cells (figure 1.6).

Cells are either defined by intersections of Regions of Space or defined by Unions of Regions of Space. MCNP treats problem geometry primarily in terms of regions or volumes bounded by first and second-degree surfaces. The table shown in figure 1.6 lists the commonly used surfaces by MCNP to create the geometry of a problem. All refer to a Cartesian coordinate system. A surface is represented functionally as $f(x, y, z) = 0$.

As far as simulations in diagnostic radiology are concerned, and since different densities were assigned for the skeletal (newborn and other ages), soft (newborn and other ages), and lung tissue, specification of materials filling the various cells in an MCNP calculation involves defining a unique material number and the elemental or the isotopic composition of this material must be stated in the MCNP input file (table 1.9).

For medical physics purposes, the specification of a source variable is usually either an explicit value of its monoenergetic energy or an energy distribution. A lot work has been done in the literature and various methods for generating X-ray spectra from diagnostic energies have been suggested. Boone and Seibert have

Table 1.9. Elemental compositions of soft and skeletal tissue used in MCNP calculations for estimating absorbed doses in paediatric and adult subjects.

Material	Density (gr cm^{-3})	% (w/w)					
Element		H	C	N	O	Ca	etc
Soft tissue (newborn)	1.04	0.106	0.149	0.017	0.718	...	0.01
Soft tissue	1.04	0.105	0.226	0.025	0.635	...	0.009
Lung tissue	0.296	0.101	0.103	0.029	0.756	...	0.011
Bone (newborn)	1.22	0.079	0.097	0.027	0.668	0.079	0.05
Bone (other ages)	1.4	0.074	0.255	0.031	0.479	0.102	0.016

suggested the use of interpolating polynomials to accurately compute X-ray spectra at 1 kV intervals ranging from 30 kV to 140 kV. This was actually a method to interpolate measured constant potential X-ray spectra of earlier studies. In other works, MCNP or another MC code by itself was used for the simulation of X-ray spectra in diagnostic radiology and mammography. In some works, the electrons were transported from the cathode to anode until they slow down and stop in the tungsten target and both bremsstrahlung and characteristic X-ray production were considered (Ay *et al* 2004). In addition, several commercial software packages are available for computing normalized or un-normalized anode spectra for X-ray diagnostic energies.

Tally cards are used to specify what type of information the user wants to gain from the Monte Carlo calculations; these are the transport code's computational detectors. MCNP offers 6 neutron tallies, 6 photon tallies and 4 electron tallies, normalized per starting particle, which can measure charge, flux and energy deposition (absorbed dose). The measured quantity indicates the type of the tally-detector and limits the application over a surface or a cell. All MCNP tally values are normalized per starting particle. This means that if the user wants to estimate dose deposition inside an organ the 'per starting particle' normalization must be 'eliminated' by implementing in the code a tally to normalize the MCNP output tally data. That tally may be inside the beam before the patient so that the organ tally value is expressed in terms of the primarily irradiated 'entrance surface dose' tally.

In order to construct the geometry of the pregnant patient or the paediatric patient, the user must either encode the geometry or use other software to implement stylized or voxelized geometry inside the MCNP input file. Visual editor is a graphical interface software to complement the MCNP distribution, which enables the user to create surfaces and cells, inspect the geometry of an input file, access the materials library, set cell importances, create sources, plot particle tracks and create cross-section and 3D plots of the input geometry. For stylized phantom calculations, software packages have been proposed for producing pregnant phantoms at the first, second and third trimester of gestation (Metger and Van Riper 1999). However, none of the 'older' stylized phantoms have organs assigned a weighting factor from

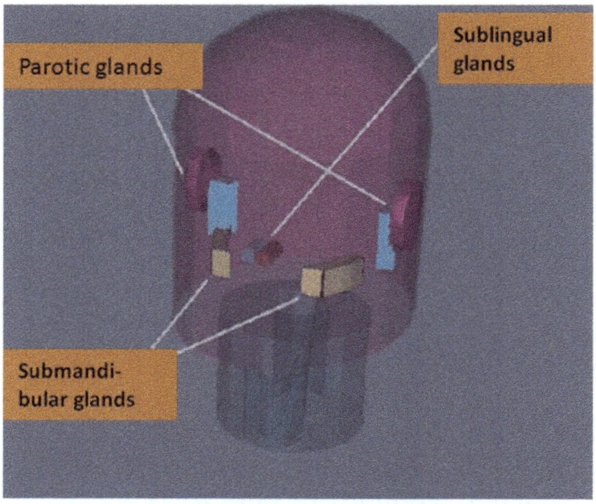

Figure 1.18. Organs not supplied in older stylized phantoms must be manually hardcoded in the phantom (MCNP Visual Editor representation).

the ICRP-103 report for the calculation of effective dose. So, organs or tissues of interest not present in older mathematical phantoms, such as salivary glands and adrenals, must be manually hardcoded in the phantom (figure 1.18).

Lymph nodes, bone and active bone marrow absorbed doses must be calculated making some assumptions. Firstly, because lymphatic nodes are not usually modeled in any phantom, the dose in the lymph nodes is estimated from the doses in surrogate organs taking into account the percentage in lymphatic tissue of small intestine, pancreas, gallbladder, thyroid, glands and other organs. The dose in the active bone marrow is calculated by the following equation as a sum over all energy depositions and all parts of the skeleton:

$$\Delta E_{\mathrm{ABM},i} = \Delta E \frac{\mu_{\mathrm{en}\left(\frac{E}{\rho}\right)\mathrm{ABM}}}{\mu_{\mathrm{en}\left(\frac{E}{\rho}\right)\mathrm{bone}}} \frac{m_{\mathrm{ABM},i}}{m_{\mathrm{bone},i}} f_i(E) \tag{1.14}$$

For an energy deposition ΔE in a specific skeletal part i from a photon with energy E, the part of energy deposited in the active bone marrow in that skeleton part, $\Delta E_{\mathrm{ABM},i}$, is calculated where $m_{\mathrm{bone},i}$ and $m_{\mathrm{ABM},i}$ denote the mass of the skeleton part i, and the mass of active bone marrow in that skeleton part, respectively. $\mu_{\mathrm{en}}(E/\rho)$ is the mass energy absorption coefficient.

Furthermore, the challenge in using MCNP and other native MC transport codes is to minimize the computing expense needed to obtain a tally estimate with acceptable relative error (as well as satisfying other statistical criteria intrinsically used inside the code). For many problems, a direct simulation (analog MCNP) would require far too many histories to achieve acceptable results with the computer time available. For such cases, the user must employ 'tricks' to reduce the relative error of a tally (or its variance) for a fixed computing time, or to reduce the computing time to achieve the same relative error.

References

Akselrod A E and Akselrod M S 2002 Correlation between OSL and the distribution of TL traps in Al_2O_3:C *Radiat. Prot. Dosim.* **100** 217–20

Akselrod M S and Gorelova E A 1993 Deep traps in highly sensitive a-Al_2O_3:C TLD crystals *Nucl. Tracks Radiat. Meas.* **21** 143–6

Akselrod M S, Kortov V S, Kravetsky D J and Gotlib V I 1990 Highly sensitive thermoluminescent anion-defect α-Al_2O_3:C single crystal detectors *Radiat. Prot. Dosim.* **33** 119–22

Akselrod M S and McKeever S W S 1999 A radiation dosimetry method using pulsed optically stimulated luminescence *Radiat. Prot. Dosim.* **81** 167–76

Alderson S W, Lanzl L H, Rollins M and Spira J 1962 An instrumented phantom system for analog computation of treatment plans *Am. J. Roentgenol.* **87** 185

Ay M R *et al* 2004 Monte Carlo simulation of x-ray spectra in diagnostic radiology and mammography using MCNP4C *Phys. Med. Biol.* **49** 4897–917

Aznar M C, Medin J, Hemdal B, Thilander Klang A, Bøtter-Jensen L and Mattsson S 2005a A Monte Carlo study of the energy dependence of Al_2O_3:C crystals for real-time *in vivo* dosimetry in mammography *Radiat. Prot. Dosim.* **114** 444–9

Aznar M C *et al* 2005b *In vivo* absorbed dose measurements in mammography using a new real-time luminescence technique *Br. J. Radiol.* **78** 328–34

Berger M J *et al* 2011 *XCOM: Photon Cross sections Database, NIST Standard Reference Database 8 (XGAM)* (Gaithersburg, MD: The National Institute of Standards and Technology (NIST))

Bilski P, Budzanowski M and Olko P 1996 A systematic evaluation of the dependence of glow curve structure on the concentration of dopants in LiF: Mg, Cu, P *Radiat. Prot. Dosim.* **65** 195–8

Blinov N N Jr. *et al* 2005 A method for determination of the effective dose received by patient during digital scanning fluorographic examination from the results of measurement of radiation dose scattered by patient's body *Biomed. Eng.* **39** 202–12

Boone J M and Chavez A E 1996 Comparison of X-ray cross sections for diagnostic and therapeutic medical physics *Med. Phys.* **23** 1997–2005

Bos A J J 2001 High sensitivity thermoluminescence dosimetry *Nucl. Instrum. Methods Phys. Res.* B **184** 3–28

Bøtter-Jensen L 1997 Luminescence techniques: instrumentation and methods *Radiat. Meas.* **27** 749–68

Bøtter-Jensen L, McKeever S W S and Wintle A G 2003 *Optically Stimulated Luminescence Dosimetry* (Amsterdam: Elsevier)

Bower M W and Hintenlang D E 1998 The characterization of a commercial MOSFET dosimeter for use in diagnostic x-ray *Health Phys.* **75** 197–204

Buffon G C 1777 Essai d'arithmetique morale, Supplement a l'Histoire Naturelle, Volume 4

Chen J 2004 Mathematical models of the embryo and fetus for use in radiological protection *Health Phys.* **86** 285–95

Cristy M 1980 Mathematical phantoms representing children of various ages for use in estimates of internal dose *Technical report* NUREG/CR-1159; ORNL/NUREG/TM-367 (Oak Ridge, TN: Oak Ridge National Laboratory)

Cristy M and Eckerman K F 1987 *Specific absorbed fractions of energy at various ages from internal photon sources, ORNL/TM-8381* (Oak Ridge, TN: Oak Ridge National Laboratory)

Damilakis J *et al* 2000 Estimation of fetal radiation dose from computed tomography scanning in late pregnancy *Inv. Radiol.* **35** 527–53

Damilakis J, Perisinakis K, Tzedakis A, Papadakis A and Karantanas A 2010 Radiation dose to the embryo from NDCT during early gestation: a method that allows for variations in maternal body size and embryo position *Radiology* **257** 483–9

Damilakis J, Tzedakis A, Perisinakis K and Papadakis A E 2010 A method of estimating embryo doses resulting from multidetector CT examinations during all stages of gestation *Med. Phys.* **37** 6411–20

Delgado A 1996 Recent improvements in LiF:Mg,Ti and LiF:Mg,Cu,P based environmental dosimetry *Radiat. Prot. Dosim.* **66** 129–34

Dimbylow P J 1996 The development of realistic voxel phantoms for electromagnetic field dosimetry *Proc. Workshop on Voxel Phantom Development (Chilton, UK)*

Dimbylow P 2005 Development of the female voxel phantom, NAOMI, and its application to calculations of induced current densities and electric fields from applied low frequency magnetic and electric fields *Phys. Med. Biol.* **50** 1047–70

Dimbylow P 2006 Development of pregnant female, hybrid voxel-mathematical models and their application to the dosimetry of applied magnetic and electric fields at 50 Hz *Phys. Med. Biol.* **51** 2383–94

Dimbylow P J 1997 FDTD calculations of the whole-body averaged SAR in an anatomically realistic voxel model of the human body from 1 MHz to 1 GHz *Phys. Med. Biol.* **42** 479–90

Dixon R L and Ballard A C 2007 Experimental validation of a versatile system of CT dosimetry using a conventional ion chamber: beyond CTDI100 *Med. Phys.* **34** 3399

Duggan L and Kron T 1999 Glow curve analysis of long-term stability of LiF:Mg,Cu,P as compared to LiF:Mg,Ti *Radiat. Prot. Dosim.* **85** 213–16

Eckerman K F and Stabin M G 2000 Electron absorbed fractions and dose conversion factors for marrow and bone by skeletal regions *Health Phys.* **78** 199–214

Edmund J M, Andersen C E, Marckmann C J, Aznar M C, Akselrod M S and Bøtter-Jensen L 2006 CW-OSL measurement protocols using optical fibre Al_2O_3:C dosemeters *Radiat. Prot. Dosim.* **119** 368–74

Evans B D, Pogatshnik G J and Chen Y 1994 Optical properties of lattice defects in a-Al_2O_3 *Nucl. Instrum. Methods Phys. Res.* **B 91** 258–62

Fill U A, Zankl M, Petoussi-Henss N, Siebert M and Regulla D 2004 Adult female voxel models of different stature and photon conversion coefficients for radiation protection *Health Phys.* **86** 253–72

Fitousi N T *et al* 2006 Patient and staff dosimetry in vertebroplasty *Spine* **31** E884

Fricke B L *et al* 2003 In-plane bismuth breast shields for pediatric CT: effects on radiation dose and image quality using experimental and clinical data *Am. J. Roentgenol.* **180** 407

Frush D P and Yoshizumi T 2006 Conventional and CT angiography in children: Dosimetry and dose comparisons *Pediatr. Radiol.* **36** 154

Giansante L, Santos J C, Umisedo N K, Terini R A and Costa P R 2018 Characterization of OSL dosimeters for use in dose assessment in computed tomography procedures *Phys. Med.* **47** 16

Gaza R and McKeever S W 2006 A real time, high-resolution optical fibre dosemeter based on optically stimulated luminescence (OSL) of KBr:Eu, for potential use during the radiotherapy of cancer *Radiat. Prot. Dosimetry* **120** 14

Hill R, Holloway L and Baldock C 2005 A dosimetric evaluation of water equivalent phantoms for kilovoltage x-ray beams *Phys. Med. Biol.* **50** N331–4

Hintenlang D, Moloney W and Winslow J 2008 An anthropomorphic adult physical phantom and fiber optic coupled point dosimetry system for the measurement of effective and average organ doses of CT patients *Med. Phys.* **35** 2652

Hirata A, Ito N, Fujiwara O, Nagaoka T and Watanabe S 2008 Conservative estimation of whole-body-averaged SARs in infants with a homogeneous and simple-shaped phantom in the GHz region *Phys. Med. Biol.* **53** 7215–23

Horowitz Y S and Moscovitch M 2013 Highlights and pitfalls of 20 years of application of computerised glow curve analysis to thermoluminescence research and dosimetry *Radiat. Prot. Dosim.* **153** 1–22

Horowitz Y S 1981 The theoretical and microdosimetric basis of thermoluminescence and applications to dosimetry *Phys. Med. Biol.* **26** 765–824

Horowitz Y S and Datz H 2011 Thermoluminescence dose response: experimental methodology, data analysis, theoretical interpretation *AIP Conf. Proc.* **1345** 187

Horowitz Y S and Moscovitch M 1986 LiF-TLD in the mGy range via computerised glow curve deconvolution and background smoothing *Radiat. Prot Dosim.* **17** 337–42

Horowitz Y S, Ben Shachar B and Yossian D 1993 Study of the long-term stability of peaks 4 and 5 in TLD-100: correlation with isothermal decay measurements at elevated temperatures *J. Phys. D: Appl. Phys.* **26** 1475–81

Hubbell J H 1982 Photon mass attenuation and energy-absorption coefficients *Int. J. Appl. Radiat. Isot.* **33** 1269–90

Hubbell J H and Seltzer S M 2004 *Tables of X-ray Mass Attenuation Coefficients and Mass Energy-Absorption Coefficients* (Gaithersburg, MD: NIST)

Hurwitz L M *et al* 2006 Radiation dose to the fetus from body MDCT during early gestation *Am. J. Roentgenol.* **186** 871

Hwang J M L, Shoup R L and Poston J W 1976 *Mathematical Description of a Newborn Human for Use in Dosimetry Calculations* ORNL TM-5453 Oak Ridge National Laboratory

ICRP Publication 89 2002 *Basic Anatomical and Physiological Data for Use in Radiological Protection Reference Values* (Oxford: Pergamon)

ICRP publication 103 2007 The 2007 Recommendations of the International Commission on Radiological Protection *Ann. ICRP* **37** 1

ICRP publication, 110 2009 *Adult Reference Computational Phantoms* (Oxford, UK: Pergamon)

ICRP Publication 23 1975 *Report on the Task Group on Reference Man* (Oxford: Pergamon Press, International Commission on Radiological Protection)

ICRU Report 46 1992 *Photon, Electron, Proton and Neutron Interaction Data for Body Tissues, International Commission on Radiation Units and Measurements* (Bethesda, MD)

ICRU Report 44 2016 *Tissue Substitutes in Radiation Dosimetry and Measurement* (Bethesda, MD)

Jones R M *et al* 1976 *The Development and Use of a Fifteen-Year-Old Equivalent Mathematical Phantom for Internal Dose Calculations* ORNL/TM-5278 OSTI.GOV

Jursinic P A 2007 Characterization of optically stimulated luminescence dosimeters, OSLDs, for clinical dosimetric measurements *Med. Phys.* **34** 4594–604

Kalos M H and Whitlock P A 1986 *Monte Carlo Methods Vol. I Basics* (New York: John Wiley and Sons)

Kramer R, Khoury H J, Vieira J W, Loureiro E C M, Lima V J M and Lima F R A *et al* 2004 All about FAX: a female adult voXel phantom for Monte Carlo calculation in radiation protection dosimetry *Phys. Med. Biol.* **49** 5203–16

Kramer R, Zankl M, Williams G and Drexler G 1982 *Part I: The Calculation of Dose from External Photon Exposures Using Reference Human pPhantoms and Monte Carlo Methods: Part I. The Male (ADAM) and Female (EVA) Adult Mathematical Phantoms* GSF-Report S-885 (Neuherberg-Muenchen: Institut fuer Strahlenschutz, GSF-Forschungszentrum fuer Umwelt und Gesundheit) Reprint July 1999

Kramer R *et al* 2006 MAX06 and FAX06: Update of two adult human phantoms for radiation protection dosimetry *Phys. Med. Biol.* **51** 3331

Kron T 1994 Thermoluminescence dosimetry and its applications in medicine–Part 1: Physics, materials and equipment *Australas. Phys. Eng. Sci. Med.* **17** 175–99

Kumar M, Kher R K, Bhatt B C and Sunta C M 2007 A comparative study of the models dealing with localized and semi-localized transitions in thermally stimulated luminescence *J. Phys. D: Appl. Phys.* **40** 5865–72

Lee C *et al* 2005 The UF series of tomographic computational phantoms of pediatric patients *Med. Phys.* **32** 3537

Lee C *et al* 2006 Whole-body voxel phantoms of paediatric patients—UF Series B *Phys. Med. Biol.* **51** 4649

Lee C, Williams J L, Lee C and Bolch W E 2005 The UF series of tomographic computational phantoms of pediatric patients *Med. Phys.* **32** 3537–48

Markey B G, Colyott L E and McKeever S W S 1995 Time-resolved optically stimulated luminescence from a-Al_2O_3:C *Radiat. Meas.* **24** 457–63

McKeever S W S, Akselrod M S and Markey B G 1996 Pulsed optically stimulated luminescence dosimetry using α-Al_2O_3:C *Radiat. Prot. Dosim.* **65** 267–72

McKeever S W S, Akselrod M S, Colyott L E, Agersnap Larsen N, Polf J C and Whitley V H 1999 Characterisation of Al_2O_3 for use in thermally and optically stimulated luminescence dosimetry *Radiat. Prot. Dosim.* **84** 163–8

McKeever S W S 1985 *Thermoluminescence of Solids* (Cambridge: Cambridge University Press)

McKeever S W S, Moscovitch M and Townsend P D 1995 *Thermoluminescence Dosimetry Materials: Properties and Uses* (Ashford, UK: Nuclear Technology Publishing)

McKeever S W S 2001 Optically stimulated luminescence dosimetry *Nucl. Instrum. Methods Phys. Res.* B **184** 29–54

McKinlay A F 1981 *Thermoluminescence Dosimetry* Bristol: Hilger in collaboration with the Hospital Physicists' Association, c1981 (Harwell: National Radiological Protection Board)

Metger R L and Van Riper K A 1999 Fetal dose assessment from invasive special procedures by Monte Carlo methods *Med. Phys.* **26** 1714–20

Mobit P, Agyingi E and Sandison G 2006 Comparison of the energy-response factor of LiF and Al2O3 in radiotherapy beams *Radiat. Prot. Dosim.* **119** 497–9

Mukundan S *et al* 2007 MOSFET dosimetry for radiation dose assessment of bismuth shielding of the eye in children *Am. J. Roentgenol.* **188** 1648

Muniz J L, Hernandez Verduzo R and Delgado A 1996 A comparison of the TLD-100 dosimetric performance using different annealing procedures and glow curve analysis *Radiat. Prot. Dosim.* **66** 273–7

Nipper J C, Williams J L and Bolch W E 2002 Creation of two tomographic voxel models of paediatric patients in the first year of life *Phys. Med. Biol.* **47** 3143

Nowotny R 1998 *XMuDat: Photon Attenuation Data on PC (version 1.0.1)* International Atomic Energy Agency (IAEA), IAEA-NDS-Documentation Series; IAEA-NDS-195 (REV.0)

Olko P, Bilski P, Budzanowski M, Molokanov A, Ochab E and Waligorski M P R 2001 Supralinearity of peak 4 and 5 in thermoluminescent lithium fluoride MTS-N (LiF:Mg,Ti) detectors at different Mg and Ti concentrations *Radiat. Meas.* **33** 807–12

Osei E K and Kotre C J 2001 Equivalent dose to the fetus from occupational exposure of pregnant *Br. J. Radiol.* **74** 629

Pagonis V, Chen R and Lawless J L 2006 Nonmonotonic dose dependence of OSL intensity due to competition during irradiation and readout *Radiat. Meas.* **41** 903–9

Park S, Lee J K and Lee C 2006 Development of a Korean adult male computational phantom for internal dosimetry calculation *Radiat. Prot. Dosim.* **121** 257

Perisinakis K *et al* 2004 Patient exposure and associated radiation risks from fluoroscopically guided vertebroplasty or kyphoplasty *Radiology* **232** 701–7

Perisinakis K *et al* 2005 Fluoroscopically guided implantation of modern cardiac resynchronization devices: radiation burden to the patient and associated risks *J. Am. Coll. Cardiol.* **46**

Perisinakis K *et al* 2006 Fluoroscopy-controlled voiding cystourethrography in infants and children: are the radiation risks trivial? *Eur. Radiol.* **16** 846–51

Perisinakis K *et al* 2016 Data and methods to assess occupational exposure to personnel involved in cardiac catheterization procedures *Phys. Med.* **32** 386–92

Piters T M and Bos A J J 1995 Dose response of thermoluminescence emission spectra of LiF:Mg, Ti with different Mg, Ti impurity concentrations *Radiat. Meas.* **24** 431–4

Sanderson D C W and Clark R J 1994 Pulsed photostimulated luminescence of alkali feldspars *Radiat.Meas.* **23** 633–9

Shi C and Xu X G 2004 Development of a 30-week-pregnant female tomographic model from computed tomography (CT) images for Monte Carlo organ dose calculations *Med. Phys.* **31** 2491–7

Snyder W S, Ford M R and Warner G G 1978 Estimates of absorbed fractions for monoenergetic photon sources uniformly distributed in various organs of a heterogeneous phantom, MIRD Pamphlet No. 5, revised, Society of Nuclear Medicine, January 1978, Society of Nuclear Medicine, New York, NY

Snyder W S *et al* 1969 *MIRD Pamphlet No. 5 Estimates of Absorbed Fractions for Monoenergetic Photon Sources Uniformly Distributed in Various Organs of a Heterogeneous Phantom* (New York: Society of Nuclear Medicine)

Somerwil A and van Kleffens H J 1977 Experience with the Alderson RANDO® phantom *Br. J. Radiol.* **50** 295

Stabin M *et al* 2008 ICRP-89 based adult and pediatric phantom series *J. Nucl. Med. Meet. Abst.* **49** 14

Stabin M, Watson E, Cristy M, Ryman J, Eckerman K, Davis J, Marshall D and Gehlen K 1995 *Mathematical Models and Specific Absorbed Fractions of Photon Energy in the Nonpregnant Adult Female and at the End of Each Trimester of Pregnancy* ORNL/TM-12907 (Oak Ridge, TN: Oak Ridge National Laboratory)

Stadtmann H, Hranitzky C and Brasik N 2006 Study of real time temperature profiles in routine TLD readout-influences of detector thickness and heating rates on glow curve shape *Radiat. Prot. Dosim.* **119** 310–13

Taylor M L, Smith R L, Dossing F and Franich R D 2012 Robust calculation of effective atomic numbers: auto-zeff software *Med. Phys.* **39** 1769–78

Theocharopoulos N *et al* 2002 Comparison of four methods for assessing patient effective dose staff in diagnostic radiology *Med. Phys.* **29** 2070

Theocharopoulos N *et al* 2006 Occupational exposure in the electrophysiology laboratory: quantifying and minimizing radiation burden *Br. J. Radiol.* **79** 644–51

Vergara Saez J C, Budzanowski M, Gomez-Ros J M and Romero A M 1999 Thermally induced fading of individual glow peaks in LiF:Mg, Cu, P at different storage temperatures *Radiat. Prot. Dosim.* **85** 269–72

Viamonte A 1, da Rosa L A, Buckley L A, Cherpak A and Cygler J E 2008 Radiotherapy dosimetry using a commercial OSL system *Med. Phys.* **35** 1261–6

Wang B, Kim C H and Xua X G 2004 Monte Carlo modeling of a high-sensitivity MOSFET dosimeter for low- and medium-energy photon sources *Med. Phys.* **31** 1003

Weizmann Y, Horowitz Y S and Oster L 1999 Investigation of the composite structure of peak 5 in the thermoluminescent glow curve of LiF: Mg, Ti (TLD-100) using optical bleaching *J. Phys. D: Appl. Phys.* **32** 2118–27

White D R 1978 Tissue substitutes in experimental radiation physics *Med. Phys.* **5** 467

Xu X G and Eckerman K F 2010 *Handbook of Anatomical Models for Radiation Dosimetry* (Boca Raton, FL: Taylor & Francis)

Xu X G *et al* 2007 A boundary-representation method for designing whole-body radiation dosimetry models: pregnant females at the ends of three gestational periods: RPI P3, P6 and P9 *Phys. Med. Biol.* **52** 7023

Xu X G, Zhang J Y and Na Y H 2008 *The Int. Conf. on Radiation Shielding-11 (Pine Mountain, GA)*

Yukihara E G and McKeever S W S 2006a Ionization density dependence of the optically and thermally stimulated luminescence from Al_2O_3:C *Radiat. Prot. Dosim.* **119** 206–17

Yukihara E G and McKeever S W S 2006b Spectroscopy and optically stimulated luminescence of Al_2O_3:C using time-resolved measurements *J. Appl. Phys.* **100** 083512

Yukihara E G, Mardirossian G, Mirzasadeghi M, Guduru S and Ahmad S 2007 Evaluation of Al_2O_3:C optically stimulated luminescence (OSL) dosimeters for passive dosimetry of high-energy photon and electron beams in radiotherapy *Med. Phys.* **35** 260–9

Yukihara E G, Whitley V H, McKeever S W S, Akselrod A E and Akselrod M S 2004 Effect of high-dose irradiation on the optically stimulated luminescence of Al_2O_3:C *Radiat. Meas.* **38** 317–30

Yukihara E G, Whitley V H, Polf J C, Klein D M, McKeever S W S, Akselrod A E and Akselrod M S 2003 The effects of deep trap population on the thermoluminescence of Al_2O_3:C *Radiat. Meas.* **37** 627–68

Zankl M *et al* 2005 GSF male and female adult voxel models representing ICRP Reference Man–the present status *Proc. of The Monte Carlo Method: Versatility Unbounded in a Dynamic Computing World, Chattanooga, TN* (La Grange Park, IL: American Nuclear Society)

Zubal I G, Harrell C R, Smith E O, Rattner Z, Gindi G and Hoffer P B 1994 Computerized three-dimensional segmented human anatomy *Med. Phys.* **21** 299–302

Chapter 2

Biological effects of exposure to ionizing radiation during gestation and childhood

Kostas Perisinakis and Virginia Tsapaki

The radiosensitivity of a human being during gestation and childhood is higher than that during adulthood. The detrimental biological effects of exposure to ionizing radiation during gestation and childhood are associated with the amount of radiation dose absorbed by human tissues and the particular features of physiological human development during these early periods of life. The latest scientific data on the observable radiation effects in humans after exposure during gestation and childhood and how this data has been used to establish risk factors for radiation-induced detrimental effects are discussed in this chapter.

2.1 Biological effects to conceptus from ionizing radiation

2.1.1 Menstrual cycle, conception and fetal development

The menstrual cycle

Menstrual cycle is the term used to describe all physiological changes occurring in the reproductive system of a female to allow oocyte release and prepare the uterus for the possibility of pregnancy. The first menstrual cycle is called menarche and occurs at puberty, that is 10–16 years of age. The menstrual cycle is repeated until the menopause, which usually occurs between 45 and 55 years of age (Reed and Carr 2015, Mihm *et al* 2011).

The menstrual cycle length refers to the number of days from the onset of menstrual bleeding (menses) of one cycle until the onset of menstrual bleeding of the next cycle. A menstrual cycle lasts, on average, for 28 days with most women having a cycle in the range of 26–35 days. It is noted that there might be variation in the length of consecutive menstrual cycles of a woman. This variation in most women is less than 4 days and these women are considered to have a 'regular cycle'. Women

doi:10.1088/978-0-7503-1317-9ch2

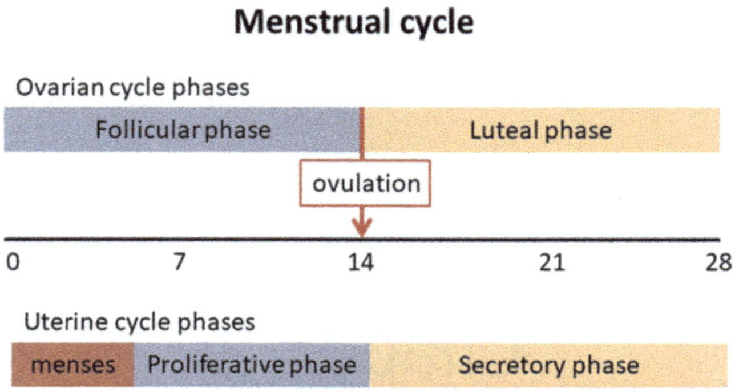

Figure 2.1. The menstrual cycle.

with variations up to 20 days are considered to have a 'moderately irregular cycle' while women with variations higher than 20 days are considered to have a 'very irregular cycle' (Reed and Carr 2015, Mihm *et al* 2011, Kippley *et al* 1996, Lenton *et al* 1984).

On the basis of events occurring in the uterus, the menstrual cycle may be divided in menstrual bleeding, proliferative phase and secretory phase. The typical duration of each phase is indicatively shown in figure 2.1. The onset of menstrual bleeding indicates the start of the menstrual cycle and the duration of menses is 2–7 days with an average of 5 days.

Regarding events occurring in the ovary, the menstrual cycle may be divided into the follicular phase, ovulation and luteal phase. During the follicular phase, the endometrial layer of the uterus is grown setting a friendly environment for a possible incoming sperm and concomitantly a primordial follicle turns into a mature follicle. Ovulation refers to the release of an oocyte by the mature follicle and usually occurs between the 10th and the 16th day after the onset of menstrual bleeding. The time of ovulation is around the 14th day in young women (age <24 years) and around the 11th day in older women.

The fertile window

The oocyte lives only for 12–24 h if not fertilized while the lifespan of sperm is about 5 days. Therefore, pregnancy may occur if a woman has sexual intercourse five days prior or one day after ovulation. This 6 day period is commonly referred as the 'fertile window' over the menstrual cycle. The fertile window is typically situated between days 10 and 17 of the average menstrual cycle (Wilcox *et al* 2000). However, even in women with a regular cycle, the fertile window may be initiated earlier or later by several days. It is noted that for a period of about 10 days after the onset of menstruation (i.e. from menses initiation to 5 days before ovulation) a woman has essentially no chance of getting pregnant.

Fertilization and conceptus development

Fertilization or conception refers to the joining/fusion of a sperm and the ovum to form a zygote cell. The formation of the zygote cell initiates prenatal development.

Meiosis occurs within the zygote cell to join genetic information carried by chromosomes of both the sperm and the ovum. This process ends up with the creation of new human individual. The zygote cell undergoes successive mitoses to form morula, which contains 16 cells and travels through the fallopian tube to the uterus where it is embedded in the uterine wall. This process is called implantation and usually occurs at about the 8th day after conception.

During the embryonic period that follows implantation, cells begin to differentiate into organs/tissues. Development of major organs, hematopoiesis and sexual differentiation occurs and therefore this stage of embryo development is also referred as 'major organogenesis'. The embryo during organogenesis is most susceptible to teratogenic environmental factors and toxic exposures. The period from 10th week after conception until birth is referred to as the fetal period. During this period all major organs/structures that have been already formed during organogenesis continue to grow and mature. The susceptibility of the fetus to deleterious effects from environmental factors is less than the embryo and toxic exposures may cause mild abnormalities or minor congenital malformations (Mole 1993, Wagner *et al* 1997).

Terminology of gestation

The time period between the first day of the last menstrual period and birth of the infant is called gestation. Gestation lasts about 40 weeks and is divided in three semesters as indicatively shown in figure 2.2. Several terms have been used in the scientific literature to describe the age of the human individual during gestation. Gestational age or menstrual age refers to the time elapsed since the first day of the last menstrual period whereas conceptional age refers to the time elapsed since fertilization/conception. It is noted that 'gestational age' differs from 'conceptional age' by about 2 weeks (figure 2.2).

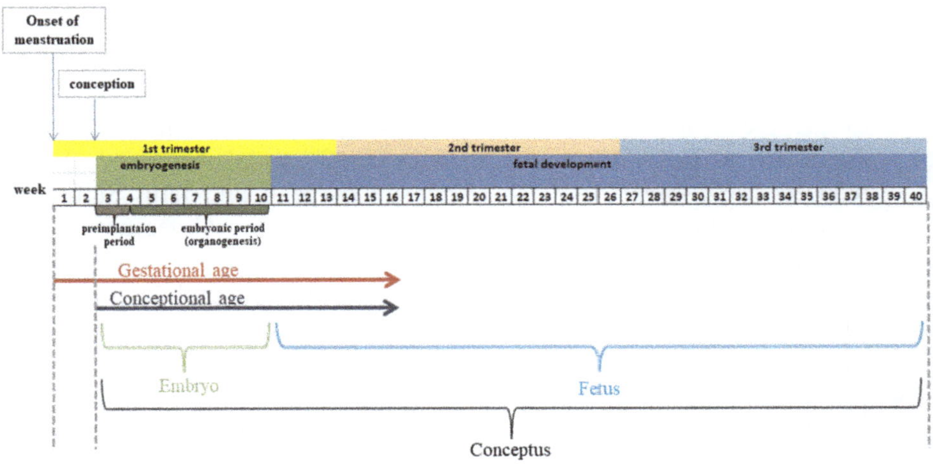

Figure 2.2. Gestation timeline.

Several terms have been used in the scientific literature to describe the human individual during prenatal development that is embryo, fetus (or foetus), conceptus, individual developing in utero and unborn child (Mole 1993, Wagner *et al* 1997).

These terms have been erroneously used interchangeably and therefore, there is confusion regarding the proper use of each term. Most scientists/researchers agree on the following use of terms (figure 2.2):

- Conceptus, individual developing in utero, unborn child (0–40th week of gestation);
- Embryo (from 2nd to 10th week of gestation);
- Fetus (from 10th week to 40th week of gestation).

2.1.2 Radiation effects of exposed conceptus

Radiation effects following in utero exposure to ionizing radiation

Exposure to ionizing radiation during gestation may have harmful effects on the individual developing in utero (ICRP 2000, 2003). The human embryo/fetus is particularly sensitive to ionizing radiation, and the deleterious effects of exposure can be severe, even for exposures that are not associated with immediate effects for the mother. In general, the radiosensitivity of the human embryo/fetus is considered to be higher than that of the human individual after birth.

Prenatal exposure of embryo/fetus to ionizing radiation may cause:

a. Lethality.

The embryo/fetus may die either prenatally (intrauterine death) or perinatally (neonatal death).

b. Growth retardation.

The embryo/fetus may not develop at a normal rate. This condition is referred as intrauterine growth restriction or retardation (IUGR) (Murki and Sharma 2014). There are a lot of reasons other than exposure to ionizing radiation that may also induce IUGR such as maternal medical disorders and pathological conditions, maternal substance abuse and toxic exposures to medications among others (Murki and Sharma 2014, ICRP 2003).

IUGR is categorized as symmetrical, asymmetrical and mixed. Fetuses with symmetrical IUGR appear with symmetrically reduced size compared to the mean size of fetus at the same gestational stage. Fetuses with asymmetrical IUGR appear with a head of normal size and a body of reduced size compared to the mean size of head and fetus at the same gestational stage. Mixed IUGR presents with clinical features of both symmetrical and asymmetrical IUGR. Fetuses with severe IUGR may die in the womb (intrauterine death) or during birth (perinatal death). Children with a less severe IUGR may have serious complications after birth such as learning disabilities, and delayed motor or social development.

c. Congenital malformations/teratogenesis.

An individual exposed in utero may present congenital anomalies of structural nature. Observed skeletal/tissue malformations include microcephaly, hydrocephaly, microphthalmia, strabismus, limb dysplasia, hupoplastic genitalia/hypospadias, among others. Such malformations have been observed in individuals exposed in utero to high levels of radiation during the atomic bombing of Hiroshima and Nagasaki (Brent 1980, Otake 1998).

d. Mental retardation.

An individual exposed in utero may present mental retardation (ICRP 2003, Otake 1998). Mental retardation (or intellectual disability that is the term currently preferred) is used to characterize the condition of humans with reduced intellectual functioning (i.e. ability to learn, solve problems, reason and make decisions) and reduced ability to communicate/interact effectively with other humans.

Intellectual ability is assessed through the intelligence quotient (IQ) measured by specific IQ tests. The average IQ of humans is considered 100 with the vast majority of individuals falling between 85 and 115. An IQ of less than 75 is evidence of intellectual disability the degree of which (i.e. mild to profound) is determined by the difference from the average IQ (i.e. 100).

e. Cancer induction.

There is evidence that an individual exposed in utero to ionizing radiation may have increased risk to develop childhood leukemia or other types of childhood cancer (ICRP 2000, 2003). Childhood cancer (or paediatric cancer) refers to the development of cancer to individuals at the age range 0–15 years. Some investigators use the term childhood cancer also for cancers developed during childhood and adolescence that is in the age range 0–19 years. The most frequent types of observed childhood cancer are leukemia, brain tumors, neuroblastoma, non-Hodgkin lymphoma, Wilms tumor, Hodgkin lymphoma, thyroid carcinoma, testicular/ovarian tumors, bone tumors and melanoma.

As recently reported, an individual born in the USA has a risk of 0.24% to develop cancer before the age of 15 years (i.e. incidence rate 24 per 10 000 individuals 0–15 years old) and a risk of 0.35% to develop cancer before the age of 20 years (i.e. incidence rate 35 per 10 000 individuals 0–19 years old) (Ward *et al* 2014). The corresponding incidence rate of childhood cancer up to 15 years of age has been recently reported for the UK to be 0.23% (Public Health England 2018). Over the last four decades, childhood cancer incidence has been reported to steadily increase (Noone *et al* 2018). Childhood cancer may be fatal or not. Incidence rate has been observed to slightly increase at an annual rate of about 0.6% over the last two decades. This upward trend may be associated with environmental changes, but more efficient diagnosis is considered the main reason for this observation (Ward *et al* 2014). In contrast, childhood cancer mortality rate has been observed to considerably decline with an annual rate of about 2.1% since 1975. This may be attributed to improved medical care and treatment.

f. Genetic effects.

Radiation exposure of animal conceptus exposed in utero has been reported to induce genomic instability that can be transmitted to the next generation where increased rate of chromosomal aberrations and structural/developmental defects may be observed (ICRP 2003). However, such data have not been confirmed in humans.

g. No deleterious effect.

The individual exposed in utero may be developed physiologically without any radiation-induced deleterious effect observed during embryo/fetal development, childhood and adolescence. It is crucial to have in mind that relatively low embryo/fetus exposures to radiation are associated with extremely low possibility for adverse radiation effects to be observed after birth and during remaining life. Therefore, in the vast majority of embryos/fetuses exposed in utero to low-level medical exposures, no harmful effect of radiation will be ever observed.

Prenatal exposure and causation of deleterious effects
All possible adverse effects to the embryo/fetus from exposure in utero may also occur in individuals not exposed to ionizing radiation prenatally. Therefore, the blame of prenatal exposure on the causation of an observed deleterious effect is occasionally impossible to be verified. For example, if an embryo is exposed prenatally to low-level radiation and develops cancer during childhood it is impossible to ascertain radiation exposure as the culprit.

Categorization of conceptus radiation effects to non-stochastic and stochastic
Radiation effects to exposed conceptus as described in the previous paragraph may be categorized in non-stochastic (or deterministic) effects and stochastic effects.

a. Non-stochastic radiation effects to conceptus.

Intrauterine death, malformations/dysplasias, growth retardation and mental retardation are considered non-stochastic radiation effects since the following apply for each of these effects:

- There is a threshold dose below which the effect is not observed.
- If a threshold dose is exceeded, the probability of induction of the effect is proportional to the amount of radiation dose absorbed by the embryo/fetus tissues.
- The severity of the effect is proportional to the amount of radiation dose absorbed by the embryo/fetus tissues.
- The effect is observed only if a high number of cells are killed/affected and therefore it may be observed only after relatively high level in utero exposures to ionizing radiation.

b. Stochastic radiation effects to conceptus.

Cancer induction and genetic effects are considered stochastic radiation effects since the following apply for each of these effects:

- There is **not** a threshold dose below which the effect is not observed and, theoretically, an absorbed dose close to zero may induce the effect.
- The effect may be observed even if a single cell is affected.

- The probability of induction for the effect is proportional to the amount of radiation dose absorbed by embryo/fetus tissues. That simply means that if an absorbed dose by the embryo/fetus tissues is close to zero, the risk for induction of the effect is also close to zero.
- The severity of the effect is **not** proportional to the amount of radiation dose absorbed by the embryo/fetus tissues.

2.1.3 Radiosensitivity of a developing conceptus

The nature, severity and possibility of occurrence of a deleterious biological effect on the developing conceptus following in utero exposure to ionizing radiation depends on the magnitude of absorbed dose to the embryo/fetal tissues and the gestational stage at which the embryo/fetus is exposed (ICRP 2003).

Conceptus exposure during the pre-implantation period

The pre-implantation period lasts about 8 days in humans with minimal duration variability between different individuals in contrast to duration of other phases of prenatal development which may vary considerably. During this phase, the single cell conceptus (zygote) proliferates through cell division to the blastocyst, which is made up by about 250 cells in humans (ICRP 2003, Kippley and Kippley 1996, Lenton *et al* 1984).

The effects of conceptus exposure to ionizing radiation during the pre-implantation period have been studied on animals. Data for humans have been extrapolated from animal studies. Radiation exposures resulting in absorbed doses above 50 mGy has been reported to affect cell proliferation during this early stage of prenatal development. Following exposures resulting in conceptus dose above this level, several cells may die resulting in death of the whole conceptus, which occurs prior to implantation. Death of the conceptus prior to implantation is normally not perceived by the few-days pregnant female, since pregnancy is not recognized yet at this early stage and there is no noticeable mild pain or other symptoms. Most investigators tend to agree that if exposure during the pre-implantation period does not result in death, the surviving conceptus develops physiologically and there is practically no risk for radiation-induced deleterious effects to be observed during prenatal development or after birth. In other words, the risk of radiation-induced cancer, structural malformation or growth/mental retardation following exposure during the pre-implantation period is unlikely. The lethal effect of exposure to ionizing radiation during the pre-implantation period is therefore considered a non-stochastic 'all-or-nothing effect' with a threshold dose of about 50 mGy (ICRP 2000, 2003).

Conceptus exposure during the period of organogenesis

The period of organogenesis lasts 40–50 days in humans following implantation. It includes the whole embryonic period after implantation (i.e. from the 5th week until the 10th week of gestational age or equivalently from the 3rd week to the 8th week post conception). During major organogenesis, the internal organs and tissues are formed and by the end of this period an embryo attains the morphological structures of a human being.

The effect of conceptus exposure to ionizing radiation during the major organogenesis period have been investigated on animals and on the atomic bomb survivors of Hiroshima and Nagasaki that were pregnant at the time of bombing. The radiation effects observed after exposure during this period include death (embryonic, fetal or neonatal), malformations of structural nature (teratogenesis), growth retardation, severe mental retardation (SMR) and reduced intelligence quotient (IQ), which are considered as non-stochastic effects and childhood cancer that is considered a stochastic effect (ICRP 2003).

If a high number of cells are killed as a result of the radiation exposure, death of the conceptus may occur. The threshold of radiation dose above which conceptus death may be induced following exposure at this period is in the range of 50–500 mGy. In general, the risk of lethality following in utero exposure decreases as the embryo development advances (ICRP 2003). A threshold dose for conceptus death about 250 mGy has been suggested for the whole duration of gestation post implantation (McCollough 2007).

If a number of cells that have started differentiating to form an organ are killed following exposure to ionizing radiation the physiological development of that organ may be obscured and a major structural abnormality may be caused. Along major organogenesis, the highest vulnerability to radiation for a specific organ/tissue is observed during the period that the differentiation process for the formation of that particular organ/tissue occurs. Threshold dose for induction of structural malformations may vary depending on the type of malformation. A threshold dose of 100 mGy, that has been reported for microcephaly in animal studies, has been suggested as the threshold dose for structural malformation, although the threshold dose for other structural abnormalities may be much higher than 100 mGy. Besides, the threshold dose for growth retardation after an exposure occurring during organogenesis has been suggested to be 250 mGy (ICRP 2003).

Animal experimental data regarding the induction of SMR following radiation exposure in utero are consistent with observations on the atomic bomb survivors. Embryo at the end of organogenesis period and up to the 15th week post conception is considerably sensitive to radiation-induced SMR with a threshold dose of 300 mGy. The IQ decline has been reported to be 25 IQ points/Gy. A threshold dose for IQ decline may not be accurately assessed but below 100 mGy the IQ decline if any, is very low to be observed.

The radiation effects on the embryo from exposure to ionizing radiation occurring during organogenesis are summarized in table 2.1 with suggested threshold doses.

Conceptus exposure during the fetal period

The fetal period extends from the 9th week post conception to delivery. During this period all major organs/structures, already formed during organogenesis, continue to grow and mature. The susceptibility of the fetus to deleterious effects of radiation is relatively less than the embryo but still exposure to radiation may cause fetal

Table 2.1. Radiation effects to conceptus following radiation exposure in utero at different gestational stages and the corresponding proposed threshold doses below which effects have not been observed (ICRP 2003).

Gestational stage	Radiation effect	Threshold dose (mGy)	Most sensitive period (w)
Pre-implantation period (0–1 w pc[a])	Lethality	50	0–1
Embryonic period (major organogenesis) (2–8 w pc)	Lethality	250	2–8
	Growth retardation	250	2–8
	Structural malformation[b]	100	2–8
	Childhood cancer[c]	—	2–8
Fetal period (9 w pc—birth)	Lethality	250	9–15
	Growth retardation	250[d]	9–15
	Structural malformation[b]	100	9–15
	Severe mental retardation	300	9–25
	Reduction of IQ	100[e]	9–25
	Childhood cancer[c]	—	9–15

[a] pc: post conception.
[b] Including microcephaly.
[c] The risk for radiation-induced childhood cancer up to the age of 15 years after radiation exposure in utero has been estimated to be 1200 per 10 000 persons each exposed to 1000 mGy, that is 0.12% per 10 mGy, while the risk for radiation-induced fatal childhood cancer up to the age of 15 years after radiation exposure in utero is considered to be 0.06 per 10 mGy (ICRP 2000).
[d] This threshold may be considered applicable to the early fetal period but increases as gestation advances.
[e] IQ has been reported to decrease by about 25 IQ points/Gy (ICRP 2003).

death, structural abnormalities, growth retardation, mental retardation and cancer induction.

The most sensitive period for fetal death, structural abnormalities, growth retardation and cancer induction is the early fetal period that is 9th to 15th week post conception while the most sensitive period for severe mental retardation and IQ decline extends from the 9th to 25th week post conception. The possibility of occurrence and severity of structural malformations decrease as pregnancy advances from early to mid and late fetal period and teratogenic effects become less prominent.

The radiation effects on the embryo from exposure to ionizing radiation occurring during the fetal period are summarized in table 2.1 with suggested threshold doses.

The risk for radiation-induced cancer following prenatal exposure

There have been several studies on associating prenatal exposure to ionizing radiation with increased risk of childhood cancer. The most famous study is the Oxford Study of Childhood Cancers (OSCC) (Stewart *et al* 1956, 1958, Stewart and Kneale 1970a, 1970b, Muirhead and Kneale 1989). This was a large epidemiological study that revealed an association between cancer during childhood and medical

X-ray examinations performed on the expectant mother during prenatal development. However, there were many studies raising methodological defects of the OSCC study (ICRP 2003). Moreover, other studies with smaller cohorts led to different results. Nevertheless, the risk factor for radiation-induced fatal childhood cancer following radiation exposure in utero is conservatively accepted to be 0.06% per 10 mGy as derived on the basis of OSCC data (ICRP 2000). The risk for childhood cancer (fatal + non-fatal) is considered to be about double; that is 0.12% per 10 mGy (Donnelly *et al* 2011). This risk factor may be translated as: if a pregnant women is exposed to ionizing radiation resulting in a conceptus dose of 10 mGy, the developing embryo/fetus after birth will have a possibility of 0.12% to develop radiation-induced cancer during childhood (i.e. up to 15 years of age) on top of the nominal possibility of 0.24% for developing childhood cancer that applies for individuals that were not exposed prenatally (Ward *et al* 2014). There might be an additional risk of cancer that will appear after childhood but there is currently no supporting data available regarding this issue.

The carcinogenetic effect of prenatal exposure to radiation may vary between different genders, gestational stage at which exposure happens and quality/dose rate of radiation exposure. Animal studies have shown increased susceptibility to cancer incidence after prenatal irradiation for females compared to males, which may be attributed to the increased incidence in cancers of the ovary, uterus and breast. Animal experiments have shown that major organogenesis periods are associated with less susceptibility to cancer induction compared to later phases of fetal development (ICRP 2003). The OSCC study, however, suggested that cancer induction following exposure in utero at the first trimester is at least as likely as in later trimesters. High quality radiation exposures and high dose rates have been associated to increased susceptibility to cancer. Nevertheless, a sole fatal childhood cancer risk factor of 0.06% per 10 mGy has been suggested for the exposed conceptus irrespective of sex, gestational age and quality of radiation (ICRP 2000).

Estimation of the probability for radiation-induced deleterious effects following in utero exposure

The probability for radiation-induced non-stochastic effects following exposure in utero may be considered zero if the resulting dose to conceptus is below the associated threshold doses. If the conceptus dose exceeds the threshold dose of a non-stochastic effect the probability of occurrence for this effect is increased with dose. Besides, the probability for radiation-induced childhood cancer that is a stochastic effect may be estimated by

$$\text{(Probability of childhhod cancer)} = \frac{0.12\%}{10\ \text{mGy}} \cdot \text{(dose to conceptus)}$$

The nominal prevalence of congenital anomalies has been reported to be about 3% of live births for the USA population and about 2% for the UK population (U.S. Department of Health and Human Services 2015, Public Health England 2017). The nominal prevalence of childhood cancer up to 15 years of age has been reported to

Table 2.2. The probability of a conceptus not to present congenital malformation and not to develop childhood cancer after in utero exposure to ionizing radiation.

Absorbed dose to conceptus[a] (mGy)	Probability of no congenital malformation (%)	Probability of no childhood cancer up to 15 years (%)[b]
0	97.0	99.8
1	97.0	99.8
5	97.0	99.7
10	97.0	99.6
20	97.0	99.5
50	97.0	99.2
100	≈97.0	98.6

[a] Above natural background of radiation exposure.
[b] Rounded values at 1st decimal.

be 0.24% and 0.23% for the USA and UK population, respectively. In table 2.2., the risk of congenital anomalies and childhood cancer for a conceptus that is not exposed prenatally and the corresponding risks for a conceptus exposed prenatally to ionizing radiation of different levels are presented under the assumption that the nominal prevalence of congenital anomalies is 3% and the nominal incidence rate of childhood cancer is 0.24%. It is noted that it is extremely rare for the conceptus to receive more than 20 mGy following a radiodiagnostic procedure performed on the expectant mother.

2.2 Biological effects to children from ionizing radiation

2.2.1 Public opinion versus true facts

It is well known that many population groups are more vulnerable to ionizing radiation than the general population. Children seem to be the most well-known group of all and a lot of concern is related to this group. Furthermore, general public perception considers children to be the most susceptible to physical, psychological, social and economic harm leading to a lot of upheaval with everything that involves them. Accidents, mistakes, medical misdiagnosis or even wrong behavior related to children can easily become front page to newspapers, television or social media for many weeks, months or years. Recent reports of radiation overexposures during computed tomography scans are continuously found in the news media and used in the wrong way. The public attention to the negative application of ionizing radiation causes often increased anxiety, especially when caring for severely ill or injured children. This anxiety may also affect the choice of strategies to address the clinical problem. Perceptions, however, greatly influence public policy. If the public is not informed accurately it usually is very difficult to develop and support effective policies and programs that promote child well-being in general and address their medical problems effectively. Public perception is the difference between the actual true fact and the most popular public opinion based on biased media reports, inaccurate or totally incorrect scientific studies or even individual experiences. A positive public perception can facilitate policy makers to design their strategic plan to address a certain issue. On the other hand, negative public perception can hinder these activities or can slow down the process. A recent report of the World Health Organization (WHO 2016) on communicating radiation risks in paediatric imaging clearly states that although imaging technology has opened new horizons for clinical diagnostics and has greatly improved patient care resulting in a wide spectrum of applications in paediatric healthcare, the knowledge and awareness of referring physicians on the health effects of ionizing radiation is low. The report concludes that physicians require sufficient background, education and accurate resources in order to have a balanced approach and make informed choices together with parents on clinical problems. Another interesting point is stated at the latest United Nations Scientific Committee on the Effects of Atomic Radiation (UNSCEAR) (UNSCEAR 2013) report on the effects of radiation exposure of children that there are no comprehensive reports that address all aspects of radiation exposure and health effects to children. For all these reasons, scientists working in the field of ionizing radiation and more specifically in the field of healthcare and medical imaging should be given all the scientifically available information to answer questions related to the subject, especially if these come from the general public.

The following sections will present a detailed review of the latest scientific literature related to the biological effects to children from ionizing radiation starting from the simple mechanisms and how these are related to anatomical and physiological characteristics of children, to the most prominent radiation-induced cancers and other radiation effects. The section will end with various imaging techniques (conventional X-ray, CT, nuclear medicine, positron emission tomography, etc) and their related cancer risks.

2.2.2 Radiation effects to children; what are the mechanisms

There are different types of radiation exposure that cause various effects. These are divided into (1) external exposure (this could be either whole body, partial body or localized exposure as shown in figure 2.3) and (2) internal exposure (such as

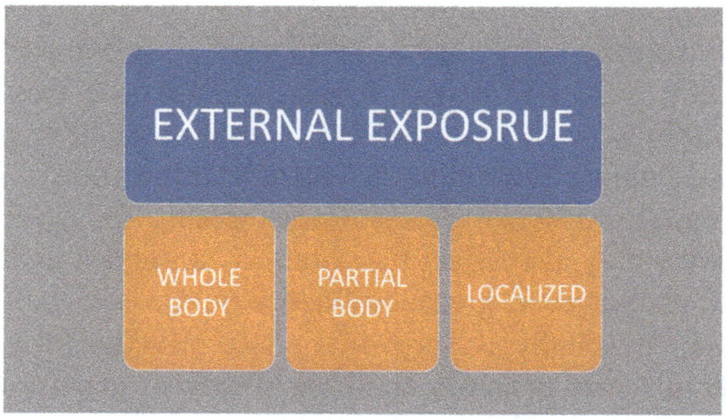

Figure 2.3. External exposure.

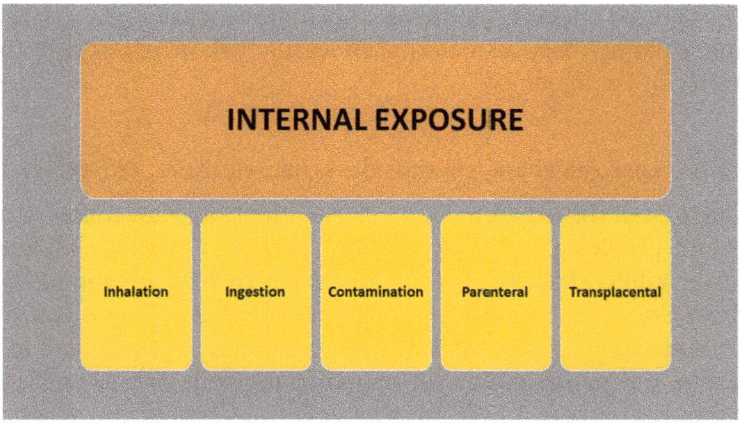

Figure 2.4. Internal exposure.

Table 2.3. Radiation effects of ionizing radiation, as these are also stated in the latest ICRP 103 report.

Early effects	Delayed or stochastic effects	
Deterministic effects/ harmful tissue reactions	Cancer	Heritable
Cell death	Due to mutation of somatic cells	Due to mutation of reproductive (germ) cells

inhalation, ingestion, contamination, parenteral or transplacental exposure as shown in figure 2.4).

Ionizing radiation exposure damages human cells. This damage can be either repaired or is irreparable. If human cell damage cannot be repaired, then the cell is either modified or the cell dies (due to damage of the deoxyribose nucleic acid strands within the chromosome). If the number of cells affected (modified or died) is high enough, organ dysfunction or death could occur.

These effects could be either early or delayed. They are classified also within the latest **ICRP 103** report and are summarized in table 2.3. As far as deterministic effects are concerned, there is always a dose threshold. Below this threshold the injury does not usually occur. Furthermore, the amount of radiation determines the severity of deterministic effect. For example, the greater the amount of radiation, the more extensive the hair loss. Deterministic effects can be seen with extensive interventional procedures, and certainly with doses delivered from radiation oncology. Deterministic effects are not encountered during diagnostic medical imaging examinations in children, except for very unusual circumstances, including imaging errors.

The stochastic effects are random effects that do not have a threshold. Also, the severity of stochastic effects do not depend on the amount of dose received.

Stochastic or delayed effects are generally disruptions that result in either cancer or heritable abnormalities. For a stochastic effect, the risk increases with the dose but not the severity of the effect.

2.2.3 Paediatric anatomical and physiological characteristics: differences from adults

Children should not be considered as small adults as they differ a lot due to anatomical and physiological changes related to age. One must understand these differences from adults and appreciate them to properly plan any radiological medical examination or assess any health effects related to ionizing radiation exposure. Organs and organ systems differ not only in the relative amount of growth and in patterns of growth but also in their positions with respect to the body and to each other.

The age ranges are set in a different way in various healthcare systems across the world. The simplest grouping is specified within the 2003 US Department of Health and Human Services Food and Drug Administration (U.S. Department of Health and Human Services 2003) recommendations:

- Infant: 0–2 years;
- Child: 1–12 years;
- Adolescent: 12–21 years;

whereas the American Academy of Paediatrics grouping is more detailed (neonatal, infant, toddler, preschool, etc).

It is very important to understand the significant anatomical and physiological differences between the different groups compared to adults (table 2.4 summarizes these differences based on the **ICRP 89** and **UNSCEAR 2013** reports).

Anatomical changes during childhood

The greatest difference between adults and children is their size.

- Head: A child's head is not only larger than that of adults, but it is also larger in relation to their whole body.
- Midpoint: The midpoint in the length of children is the umbilicus while in adults it is the symphysis pubis.
- Skin: Children's skin is much thinner and their epidermis is under-keratinized compared to that of adults. This characteristic makes them more susceptible in the absorption of harmful agents.
- Body surface: The body surface area to total body weight is larger than that of adults. As children grow up and reach adolescence these proportions change and the physical appearance becomes more like that of adults.
- Body weight: As children grow up, changes in body weight and height follow specific patterns. Within ten days after a child's birth, a loss of weight is observed. After this loss of weight is regained, the infant gradually begins gaining weight. For the first three months the infant gains weight gradually and when it reaches the age of five months its weight is twice the weight of its birth. From six months old the infant gains about one kilo per month and when it reaches one year old, its weight is three times higher than the weight

Table 2.4. Differences between anatomical and physiological characteristics of children and adults are summarized below.

System/organ or function	Children	Adults	Comments
Head and brain	Larger than adults and larger than whole body. 10% of total body weight	2% of total body weight	
Midpoint	Umbilicus	Symphysis pubis	
Skin	It is much thinner and their epidermis is under-keratinized compared to adults		
Breast	Gradually develop when ovaries start to secrete eostrogen		Density and amount of glandular tissue change during menstruation with varying oestrogen and progesterone levels for both groups
Body surface	Body surface area to total body weight is larger than adults		
Weight	1–10 days loss of weight. 1 y: 3 times weight at birth. 2 y: 4 times weight at birth. 5 y: 6 times weight at birth. 10 y: 10 times weight at birth		
Height	2–14 y: height = age (years) × 2.5 + 30		
Skull	32% of total skeletal weight at birth	12%	
Skeleton	More of cartilage than bone. Bones are less dense and more porous		Cartilage is quite radioresistant
Bone marrow	1.3% of total body weight in newborn	4.5% in adults	Radiosensitive
Neurological system	15% at birth	2%–2.5%	No major difference in mucociliary transport velocities or clearance rates
Respiratory system	Number of alveoli: 20-fold increase between birth and adulthood. Trachea is smaller and shorter		Breathing is more difficult for children than adults. Children are more vulnerable to respiratory infection

(*Continued*)

Table 2.4. (*Continued*)

System/organ or function	Children	Adults	Comments
			and to hostile ventilatory effects of anesthesia
Cardiovascular system	Heart mass/total body mass is higher than adults. Myocardium is less contractile. Cardiac output at birth is about 600 ml min^{-1}. Heart rate 150 beats/min. At birth, systolic blood pressure is approximately 80 mmHg	Cardiac output: 6000 ml min^{-1}. Heart rate 80 beats/min. Blood pressure about 120 mmHg	
Gastrointestinal system	Liver is large. Liver and spleen are located lower and more anterior; thus not protected by the rib cage		Absorption of some elements and radionuclides is greater in infants than children and adults
Genitourinary system	Testes volume is less than 5 ml up to 12 y. Ovaries volume is 2 ml up to 8 y. Bladder capacity is about 100 ml and is located in the abdomen. As the child grows, bladder moves to the pelvic region	Testes volume is about 31 ml. Ovaries volume more than 6 ml. Bladder capacity about 400–500 ml	
Endocrine system	Metabolic rate of radioiodine may be higher in children		Thyroid gland is the most susceptible organ to radiation-induced cancer
Physiological functions	Require 3–4 times more energy. Consume twice oxygen/kg. 70%–80% water in body. Higher fluid requirements. Higher need for glucose	50%–60% water in body	Children are thus more susceptible to contaminants in food or water; have greater risk for increased loss of water when ill or stressed

of its birth. At the end of its second year its weight is quadrupled. After the second year the changes in weight are more irregular and cannot be predicted. When reaching the age of five years, its weight is six times higher than the birth weight and when ten years old its weight is ten times the birth weight.
- Body height: Changes in height follow similar patterns as with weight as children grow up. Length is doubled by the end of the 4th year and tripled by the 13th year.
- Skeletal system: There are several anatomic differences in the skeletal system between children and adults.

More specifically:

a. Skull:

The skull comprises 32% of total skeletal weight at birth and has two fontanels, the anterior and the posterior. The posterior has fused when the infant reaches three months old while the posterior fuses between 10 and 16 months. When the child reaches adulthood, the skull consists of only 12% of skeletal weight.

b. Skeleton:

The skeleton of young children consists more of cartilage than bone and as children grow up the proportion of cartilaginous tissue decreases. Initially, an infant's spinal column has only the anterior curvature. When an infant begins to hold up their head and walk, then cervical and lumbar curvature appear. Ossification of cartilage takes place in ossification centers located in the center of developing bones. This procedure begins before birth in the primary ossification center. When an infant is born, most bones have not only a primary but also some secondary ossification centers. The secondary centers (epiphyses) are located at the end of long bones, they grow towards the bone with age and finally fuse during adolescence. Longitudinal growth of bones finishes at 18 years in females and 20 in males. In addition, at birth the sacrum is not fused, and the spinal column has not formed all the curvatures of the adult.

c. Long bones:

Long bones in children consist of three different main regions, which are called epiphysis, diaphysis and metaphysis. In infants and young children there is more vascularity and remodeling of bones. Bones in children are less dense, more porous than these of adults and as a consequence fracture patterns and healing mechanisms are different too.

d. Bone marrow:

In the interior of mainly long flat bones there is a flexible, spongy, nutrient-rich tissue called bone marrow. More specifically, the bone marrow is located in the trabecular bone cavities of long tubular bones and the vertebra, ribs, sternum and the flat bones of the skull and pelvis. There are two kinds of bone marrow: red and yellow. The role of red bone marrow is to help in the production of cellular blood components as it contains haematopoietic stem cells in contrast to yellow marrow that is mostly fat and does not have an important function. The total amount of bone marrow increases from 1.3% of total body weight in newborn infants to 4.5% in adults. Although there is not a big difference in the percentage of active bone marrow in total body weight between infants and adults (2%–3%) there is a change in its distribution within the bone. At birth, all the marrow is red and with age a part of red marrow transforms to yellow and the red marrow moves from the peripheral skeleton to more central portions. As mentioned in the **UNSCEAR 2013** report, bone marrow is very important due to well documented types of radiation-induced leukemia.

e. Neurological system:
   - Brain.
     The brain of the infant is quite large and it constitutes 10% of total body weight, while in adults this percentage falls to 2%. Its development needs 10 years to be completed, as in birth only 25% of the neuronal cells that exist in an adult's brain are present.
   - Blood-brain barrier.
     The blood-brain barrier is not mature yet and cerebral vessels, the network of blood vessels that help the blood circulation for brain supply, are fragile and have thin walls, especially in premature infants. Drugs such as barbiturates, opioids, antibiotics and bilirubin cross the blood-brain barrier easily causing a prolonged and variable duration of action.

f. Autonomic–parasympathetic system:
   While at birth the parasympathetic system is intact and functions properly, the autonomic system is not mature yet. The development of glia, synapses, axons and myelin around the axes is necessary for the maturation of the central nervous system. During the first two years of a child's life, axon diameters and myelin grow up significantly. When a child reaches their 3rd year of age myelination finishes. In children less than one year old, many synapses are produced and are connected by a procedure named 'pruning', which is based on experience. Gray matter remains high until the age of 12 or 16 when the level starts declining until it reaches the adult level. In general, neuronal development finishes at the age of 12 but the nervous system is fully complete only when a person reaches adulthood.

g. Cardiovascular system.
   The cardiovascular system includes the heart, blood and blood vessels. There are many profound changes after birth in the cardiovascular system. In children, the proportion of the heart's mass in relation to total body mass is high and reduces as the child grows. The heart weight increases with age and by the first year it has doubled while by the 9th year it has grown six-fold. In addition, as the child grows up, cardiac blood vessels increase in number in order to be able to supply cardiac muscle fibers that increase. In neonates the myocardium is less contractile causing the ventricles to be less compliant and less able to generate tension during contraction. This limits

the size of the stroke volume. Despite all these changes, the stroke volume is relatively fixed. The patent ductus contracts in the first few days of life and will fibrose within 2–4 weeks. Closure of the foramen ovale is pressure dependent and closes in the first day of life but it may reopen within the next 5 years.

h. Respiratory system:

The respiratory system consists of the nose, mouth, pharynx, larynx, the airways of the trachea, bronchi and bronchioles and the alveoli. Their tongue is large; larger in relation to the size of the adult's oral cavity and occupies the entire oropharynx. They have a long and stiff epiglottis that is U-shaped. Their nasal passages are narrow and the pharynx is soft tissue.

- Larynx.

  A child's larynx is anterior and is located high at the level of C3–C4, in order to allow both breathing and swallowing simultaneously. Consequently, young children breathe almost entirely through their nose. The trachea is short, and the airway is funnel shaped and narrower than that of adults. While in adults the narrowest point is at the level of vocal cords, in children the narrowest point is at the level of cricoid cartilage, where the epithelium is not tightly bound to the underlying tissue. The cartilaginous rings are quite pliable, and this results in a high risk of collapse with high inspiratory pressures. Generally, the airway of children is not uniform but consists of different heterogeneous entities, though the largest proportion is soft tissue.

- Lungs.

  In children, the chest wall is compliant and the orientation of the ribs changes as the child grows up. At birth, the orientation of the ribs is horizontal and by the time the child reaches their 10th year of age the orientation is downwards.

  While the lungs develop until the 2nd year of life, they remain immature until the age of 8 years. There are few data about the inner diameter of bronchi. There are indications that in infants the inner diameter is about 1 mm and increases to 1.5–2 mm until the 5th year and becomes 4 mm in adults. There are no great differences in the cilia or epithelial height with age. Until the age of five, the diameters of peripheral airways are narrower than those of central airways and this results in increased resistance in comparison with adults.

  The number and size of alveoli increase significantly until the child becomes 2 years old and then this increase continues more gradually until the 8th year. A full term infant, at birth, has 20–50 million alveoli and by the 8th year this number has increased to 400 million. From the 8th year and until adulthood the number stays stable and further growth is observed only in their size. The alveolar surface area is 2.8 m^2 at birth, has become 32 m^2 at 8 years old and by adulthood the surface has reached 75 m^2. These processes increase the air-tissue interface between infants and adults.

i. Alimentary tract or gastrointestinal (GI) system:

The gastrointestinal tract (GI tract or GIT) is a system responsible for transporting and digesting foodstuffs, absorbing nutrients, and expelling waste. It consists of the mouth, tongue, salivary glands, pharynx, oesophagus, stomach, intestines, liver, gallbladder and pancreas and is divided into the upper and lower GI tracts. The upper portion of the system possesses some respiratory features, since the mouth and parts of the pharynx have shared alimentary and respiratory pathways. It includes the main organs of digestion, namely, the stomach, small intestine, and large intestine, together with the accessory organs of digestion such as the tongue, salivary glands, pancreas, liver and gallbladder. In newborn children, the liver is large and during childhood, liver and spleen are located lower and more anterior. As a consequence these organs are not protected by the rib cage. Children have weaker abdominal muscles, which give an appearance of abdominal distension. As already mentioned at the respiratory system section, there are several anatomic differences between adults and children in the oral cavity, mouth and pharynx. Children have a larger tongue than adults. The epiglottis of children is floppy and omega shaped and is located at level of C3–C4 vertebra while in adults it is firm and flatter and located at level of C5–C6 vertebra. The trachea is smaller and shorter than that of adults. The larynx of children is funnel shaped with the angles posteriorly away from glottis and when they reach adulthood it has become column shaped and is positioned straight up and down. The narrowest point in children is the subglottic region but when they grow up the narrowest point is at the level of the vocal cords. The infant when born feeds exclusively on milk. The transition from liquid to textured diet starts when the infant reaches the age of 6 months, though milk still remains an important component of diet. The transit time of material in the colon is increased as the infant grows up.

j. Genitourinary system:

The genitourinary system consists of the kidneys, urethras, bladder, uterus, vagina, ovaries, prostate and penis.

- Kidneys.

  When the child is born, both kidneys are already well formed and contain the maximum number of glomeruli, though they are much smaller than those of adults. The bladder is located to the abdomen and as the child grows up it moves to the pelvis.

- Testes.

  The testes have a dual role: to produce hormones and sperm. From birth to adulthood many changes are observed but the most obvious is the increase of testes' volume; for ages less than 12 years old the volume is less than 5 ml and reaches the adult size (31 ml) by the age of 17. The hormone produced by the testes—testosterone—is produced by Leydig cells, which appear in puberty in the interstitium.

- Ovaries.

  In female children ovaries are small, with a volume less than 2 ml before the 8th year but during adolescence they become larger and more vascular. By the age of 15 ovarian volume has reached 6 ml. Also, uterine volume, which is less than 3 ml before the 8th year, becomes 15–20 ml when the child reaches 13 years old. At birth, the ovaries contain 0.3–1 million follicles, which is the greatest number of follicles, and as the child grows up this number decreases due to apoptosis. Some follicles will develop and become primary and then secondary oocytes during the menstrual cycle. In addition, ovaries produce hormones, which are called oestregen and progesterone.

k. Endocrine system:

The endocrine system includes the glands that secrete hormones in to the circulating system to be carried to certain organs. These glands are the thymus, pineal gland, pituitary gland, pancreas, ovaries, testes, thyroid gland, parathyroid gland, hypothalamus and adrenal glands (li-endocrine glands). There is a rise in the thyroid stimulating hormone (TSH) at birth that declines to normal levels within the first five days. TSH, triiodothyronine and thyroxine serum values are all highest during the first year and then decrease by about 20%–40% by the age of 16–20 y.

Physiological changes during childhood

The intense processes of growth and development in children result in higher basal metabolic rates than adults. Children need much more energy, which can be approximately 3–4 times more than those adults need. They also consume more oxygen than adults—twice the oxygen per kg. Because of this high metabolic rate they are more susceptible to contaminants in food or water and have greater risk for increased loss of water when ill or stressed.

In children, a big proportion of body weight is water. While this percentage in adults is 50%–60%, in infants the percentage of water to total body weight reaches 70%–80%. In addition, children have a larger body surface area than the adults. As mentioned already, small children have a big ratio of skin to size that reduces as the child grows up. Thus, children lose more fluids and have high fluid requirements, much higher than adults. Loss of fluids is greater when children are ill or stressed. Illness or stress accelerate the metabolic rate. Attention is needed because this can lead to respiratory failure and shock. Other differences include total circulating blood volume (80–90 ml per kg), which is 25% per unit of body weight higher than that of adults. Also special attention is needed, as immune systems in children are immature and there is a higher risk of infection in children than in adults.

a. Cardiovascular system:

Children's resting cardiac output is high in order to meet oxygen demands but decreases as the child grows up and becomes an adult. More specifically, at birth, cardiac output is about 600 ml min^{-1}, and by adolescence it has reached more than 6000 ml min^{-1}. Although the cardiac output is high, the ability of a newborn child to increase its output during stress is limited. In order to maintain the increased cardiac output at these high levels, an increase in the heart rate is necessary. Thus, normal heart rate, which is approximately 150 beats/minute, gradually decreases as the child grows up.

- Blood volume.

  Blood volume is on average 86 ml kg^{-1} in neonates and 106 ml kg^{-1} in premature infants while in normal adults it is approximately 70–80 ml kg^{-1}.
- Blood pressure.

  Blood pressure is lower than that of adults and increases as the child grows up. At birth, systolic blood pressure is approximately 80 mmHg and at puberty reaches 120 mmHg. Diastolic blood pressure also increases, as myocardial mass increases too, in order to ensure adequate coronary blood flow during diastole.

b. Respiratory system:

The airways of children have a small diameter and this results in an increased resistance to airflow. In infants, the airway and the chest wall are very compliant. The surrounding tissues poorly support the airway and the ribs that are horizontally located cannot support the chest wall. As a result, negative intrathoracic pressure is difficult to be maintained and breathing becomes more difficult for children than adults. Respiratory rates are higher in children than in adults. This is due to the high metabolic rate and oxygen consumption, which in neonates is twice than that of adults. As tidal volume remains stable, increased ventilation leads to an increase of respiratory rate. In neonates it is 30/min and falls to adult values by adolescence. Finally, children's respiratory system muscles have great oxygen and metabolite demands. The work of breathing can account for up to 40% of the cardiac output, particularly when stressed.

The changes in lung mechanics compared to those of adults increase the vulnerability of infants to respiratory infection. The same occurs to the hostile ventilatory effects of anesthesia. Secretions resulting from cholinergic activity may cause respiratory difficulty. This includes breath holding and coughing on induction of inhalational anesthesia, as well as an increased incidence of bronchospasm and even more significantly, laryngospasm. Finally, children may be more susceptible to agents absorbed through the pulmonary route than adults with the same exposure. They may also respond more rapidly to such agents. Signs and symptoms in children may be an 'early warning' of a chemical, biological, or radiological incident.

c. Gastrointestinal (GI) system:

In infants, lower esophageal sphincter tone is decreased. This fact, in conjunction with the inability to co-ordinate breathing and swallowing, results in gastroesophageal reflux, which lasts until the age of 5 months.

Transit times are different in the oesophagus, stomach and small intestine for liquids and solids. In the oesophagus, transit time for liquids is a few seconds and approximately 1 h in the stomach while for solids it is almost 100 min in the stomach. In the small intestine and colon, transit time increases with age and reaches 12–16 h in adulthood. Some elements such as calcium, lead and iron are absorbed more in childhood than in adults.

The liver function is not initially matured. Its function is decreased as enzyme systems of the hepatic system exist in infants, but they still have not sensitized or induced.

The stored fats in neonates are limited and gluconeogenesis is deficient. Finally, plasma proteins are lower than that of adults.

d. Neurological system:

When a child is born, its nervous system is anatomically complete. However, myelination is functionally immature. It continues rapidly until the age of 2 years old and is completed when the child reaches 7 years old. The blood-brain barrier continues developing for 3 years after birth. The only energy source for the brain is glucose as it is the only molecule able to cross blood-brain barrier. Although children need enormous quantities of glucose in comparison with adults, glucose is not stored in the brain and glycogen is not elaborated.

e. Skeletal system:

The bone matrix consists of organic components as well as inorganic salts. It contains carbohydrates, lipids and proteins, though the most basic component is collagen (protein). The role of collagen fibers is to be aligned along lines of stress to provide support. Inorganic salts (hydroxyapatites) such as calcium and phosphate are found in the matrix as well as in the collagen. These hydroxyapatite crystals are able to conjugate with ions such as lead, sodium, fluoride, potassium, carbon and magnesium.

As children grow up there are changes in the vascularity and porosity of bones. In infants, bones have more vascularity and remodeling than those of young adults. Also, there are slight differences in the chemical composition of bones as the child grows up. For example, the bone contains 17% by weight calcium and 23% ash at the first year of life and 21.5% and 26% at 35 years old, respectively. Bones of children have a lower modulus of elasticity, lower bending strength and as mentioned above, lower mineral content. These facts play an important role in the case of fractures.

f. Genitourinary system:

Though the bladder capacity of an adult is 400–500 ml, its capacity in infants is approximately 100 ml. When the child reaches the age of two, the capacity has increased to 175 ml, at the age of five to 250 ml and at the age of ten it is about 350 ml. Children void every 2–3 h though there are indications that most children void before bladder capacity is reached. This inability to concentrate urine is due to tubular immaturity. At birth, the kidneys are not mature and renal vascular resistance is high. Consequently, glomerular filtration and tubular function are reduced. Tubular function remains immature until 8 months. All these result in a much lower value of glomerular filtration rate at birth, a sharp rise during the first week and by the age of 2 years it reaches adult rates as renal function is now mature.

Finally, as mentioned above, in children a big proportion of body weight is water, more than in adults. In addition, 40% of body weight in children is extracellular fluid in contrast to adults where this percentage is 20%. As a result, children need large amounts of water for organ perfusion and metabolism and there is a great risk for dehydration, which is poorly tolerated by infants.

g. Endocrine system:

As ready mentioned, the thyroid gland, as well as the pituitary gland, are parts of the endocrine system. The thyroid gland accumulates iodine from the blood stream as part of its normal metabolism. It is one of the most susceptible organs for cancer induction by radiation. The pituitary gland secretes TSH that stimulates the thyroid gland to produce the two thyroid hormones called thyroxine and triiodothyronine. Growth hormones are hormones that stimulate growth, cell reproduction and cell regeneration. An example is somatotropin. It is secreted by the anterior pituitary gland and its levels rise steadily through puberty. Its main effect is to increase body height, though it has also other effects, such as increase of calcium retention, of muscle mass, stimulation of the immune system as well as stimulation of growth of all organs except the brain. During puberty, the hypothalamus of the brain begins releasing gonadotropin-releasing hormone. As response, luteinizing hormone and follicle-stimulating hormone are secreted by the anterior pituitary. This procedure results in the growth of testes and ovaries that begin producing testosterone and oestradiol. In boys, a part of the testosterone secreted is converted to oestradiol and bone growth, maturation and epiphyseal closure begin. In girls, the increased levels of oestradiol result in bone changes, breast, uterus and endometrium growth.

2.2.4 Most prominent radiation-induced cancers in children

Tumour induction in children after radiation exposure is quite variable and depends on many factors such as tumour type, age and sex. It is encouraging to know that from all tumour types, only for approximately 25% of those, children are clearly more radiosensitive than adults. Current data show that exposure in childhood increases the risk of leukemia, breast, thyroid, skin and brain cancer. Age dependence for these cancers, which are among the diseases most readily induced by radiation, is complex.

Leukemia

Leukemia was the first cancer to be linked with radiation exposure in A-bomb survivors and has the highest relative risk of any cancer. Acute forms of leukemia predominate and occur more rapidly after exposure than chronic granulocytic leukemia. Studies of survivors of atomic bomb explosions showed an increased risk in both incidence of leukemia and associated mortality. Furthermore, the risk of leukemia from radiation is higher than for other risk factors and occurs earlier than for solid cancers. Leukemia risk is best described by non-linear fit; risk is higher for exposures that occur in childhood but tends to begin to decrease 10–15 years after exposure. A linear quadratic dose response seems to provide the best fit to data. Most of the studies of radiotherapy and diagnostic irradiation confirm an increase in leukemia risk at high doses. The observed association between childhood leukemia and in utero exposure to diagnostic X-rays has been interpreted as providing further support for the etiological role of ionizing radiation, despite methodological limitations of some of the studies. Most types of radiation-induced leukemia have a minimum latency of about 2 years.

Breast cancer
Breast cancer risk was associated with radiation exposure in the A-bomb survivors cohort and among several medically exposed groups. The risk increases linearly with radiation dose and is particularly high for those exposed at young ages. The risk of breast cancer was increased in women who were under 10 years of age at the time of the atomic bomb explosion—a time when girls have little or no breast tissue. Women who developed early-onset breast cancer might have been genetically susceptible to radiation. Similar results were observed in women exposed to ionizing radiation due to repeated fluoroscopy. Age at exposure strongly influenced the risk of radiation-induced breast cancer with young women being at highest risk.

Thyroid cancer
Thyroid gland tissue is highly susceptible to radiation during childhood. Like breast cancer, thyroid cancer risks are described well by a linear dose-response function and also show a strong dependence on age at exposure. In the LSS cohort, a significant association was found between radiation dose and risk of thyroid cancer for those exposed before 19 years of age. Irradiation in childhood for benign conditions, as well as therapeutic exposure, can increase the risk of thyroid cancer. Risk is highest for children and decreases with increased age at exposure. The excess risk can be observed for many years after exposure and is highest 15–30 years after exposure. The most dramatic finding after the Chernobyl incident was a large increase in thyroid cancers in children. Approximately 2000 children have now been diagnosed with thyroid cancer, and in some locations where contamination was highest, such as in the Gomel region, the incidence of this cancer increased over 100-fold. According to the latest **IAEA 2006** report, from 1992 to 2002 in Belarus, Russia and Ukraine, more than 4000 cases of thyroid cancer were diagnosed among those who were children and adolescents at the time of the accident, the age group 0–14 years being most affected.

Brain cancer
Ionizing radiation is related to brain tumours, although the relationship is weaker than for the cancers described above. Most brain tumours associated with ionizing

radiation are benign. Japanese data show no association with brain cancers, but an increase in malignant brain tumours has been observed in patients who received radiotherapy. The evidence is strongest for those exposed before 20 years of age. In general, children exposed to radiation at ages below 20 y are approximately twice as likely to develop brain cancer compared to adults for the same radiation exposure.

2.2.5 Diagnostic imaging and cancer risk: latest data

Medical imaging is extremely important since very often they are the only tests that can help diagnose certain illnesses, aid decision-making on the best treatment and help patients avoid undergoing several other tests or even surgery. There are many types of medical imaging procedures that are used on children, each of which uses different technologies and techniques. Computed tomography (CT), conventional radiological imaging, ultrasound (US) and magnetic resonance imaging (MRI). Imaging has to take into account the dynamics of a growing body, from pre-term infants to large adolescents, where the organs follow growth patterns and phases. Due to the differences of anatomy and physiology of children and due to their delicate psychological conditions when in hospital, imaging should be done in a children's hospital, which has all the facilities necessary to image and treat, if needed, children and their specific pathologies.

Paediatric imaging is focused on conventional plain radiography and US. However, CT is increasing as it is proven to be necessary in support of routine paediatric care pathways. According to the **UNSCEAR** 2013 report, a radiation dose received by children and adults from the same source of ionizing radiation can have differing impacts. They should be considered separately in order to predict more accurately the risk following exposure for a given radiation dose and this risk is not always immediate but extends later into life.

Of all ionizing radiation imaging CT is the most important. It is widely used to help diagnose a wide range of conditions due to injury or illness and it is a valuable tool in this respect. Especially after the emergence of multi-detector row CT, the role of CT in children has been greatly extended. This has altered the way in which data is acquired. CT is now perceived as the 'gold standard' in the detection of certain pathologies. The range of available post-processing tools provide alternative ways in which CT images can be manipulated for review and interpretation in order to enhance diagnostic accuracy.

As far as general clinical problems in children, CT is typically used to diagnose:

- causes of abdominal pain;
- evaluate for injury after trauma;
- diagnose and monitor infectious or inflammatory disorders;
- diagnose and stage cancer;
- monitor response to treatment for cancer;
- evaluate blood vessels throughout the body.

Table 2.5. Summary of various CT examination types and clinical indications for children.

CT exam type	Clinical indication
Chest	Complications from infections such as pneumonia
	A tumor that arises in the lung or has spread there from a distant site
	Airway disease such as inflammation of the bronchi (breathing passages)
	Birth defects
	Pectus deformities
	Scoliosis evaluation
	Trauma to blood vessels or lung
Abdomen	Bone trauma
	Stone evaluation
	Urinary tract calculi
	Diagnose appendicitis
	Detect abdominal tumors or birth defects
Pelvis	Evaluate for stones in the urinary tract
	Assess disease of the pelvic bones
	Detect cysts or tumors in the pelvis
Head	Acute head trauma
	Craniosynostosis/plagiocephaly
	Calvarial bone lesions (Langerhans cell histiocytosis, neuroblastoma, etc)
	Suspected acute intracranial hemorrhage
	Immediate postoperative evaluation following brain surgery (evacuation of hematoma, abscess drainage, etc)
	Suspected shunt malfunctions, or shunt revisions if rapid brain MRI is not available
	Increased intracranial pressure
	Acute neurologic deficits
	Suspected acute hydrocephalus
	Brain herniation
	Suspected mass or tumor
	Non febrile seizures
	Detection of calcification

The main examination types of CT related to clinical indication are found in table 2.5.

Latest data on cancer risk and diagnostic imaging

Regarding low doses of radiation and their ability of inducing cancer in humans, evidence is limited for doses lower than 100 mSv. At the doses corresponding to a few CT scans there are direct epidemiological data from about 30 000 A-bomb survivors who were on the peripheries of Hiroshima and Nagasaki, and who were exposed in this low-dose range. This low-dose subpopulation has been followed for more than 60 years and shows a small but statistically-significant increased cancer risk.

The most recent comprehensive reports that provide either estimates of risk related to low radiation dose and/or literature review are: (1) **BEIR VII 2006** report and (2) **UNSCEAR 2013** report, which is focused on children.

BEIR VII Report 2006: The Biological Effects of Ionizing Radiation (BEIR) Committee is a group of experts in the USA with the task to develop the best possible risk estimate for exposure to low-dose, low-LET radiation in human subjects. The BEIR VII report is the seventh in a series of publications.

The main conclusions of the report are the following:

1. The BEIR VII committee recommends that in the interest of radiological protection, there be follow-up studies of cohorts of persons receiving CT scans, especially children. In addition, the committee recommends studies of infants who experience diagnostic radiation exposure related to cardiac catheterization and of premature infants who are monitored with repeated X-rays for pulmonary development.
2. The BEIR VII committee decided that the linear no-threshold model (LNT) provided the most reasonable description of the relation between low-dose exposure to ionizing radiation and the incidence of solid cancers that are induced by ionizing radiation.

UNSCEAR 2013 report: This document presents the latest data on the risks of ionizing radiation in childhood. The data come from studies on medical diagnostic radiology, radiotherapy, accidents, atomic bombings in Japan, fallout from nuclear weapon testing and potential radioactive discharges from nuclear installations. Specifically, for the atomic bombings in Japan, the survivors received relatively low doses and the follow-up has been carried out for more than 60 years. The document contains useful data on thyroid cancer from studies carried out in children exposed during the Chernobyl accident. The limitation of these studies is the shorter period of follow-up, which is 25 years, to derive accurate results on long term risks.

European projects

- The 'Epidemiological study to quantify risks for paediatric computerized tomography and to optimise doses' (**EPI-CT**) European study is an ongoing European project that started in 2010 with 18 participating centers from various European countries. It was set up to investigate the relationship between the exposure to ionizing radiation from CT scans in childhood and adolescence and possibly attributable late health effects. The centers will cooperate to enroll approximately one million patients. As reported by Thierry-Chef *et al* (2013), data collection is split into two time periods, before and after introduction of the Picture Archiving Communication System (PACS). Prior to PACS, a multi-level approach was developed to retrieve information from a questionnaire, surveys, scientific publications and expert interviews. After the introduction of PACS, scanner settings will be extracted from the Digital Imaging and Communications in Medicine headers. Radiation fields and X-ray interactions within the body will be simulated using phantoms of various ages and Monte-Carlo-based radiation transport calculations. Individual organ doses will be estimated for each child using an accepted calculation strategy, scanner settings and the radiation transport calculations. The knowledge gained on current and past CT examination practice will help to propose strategies for further dose reduction. Results were not published at the time of this report.

- The **MEDIRAD** European Project is a 4 year project aimed at enhancing the scientific bases and clinical practice of radiation protection in the medical field. One of the workpackages is entitled 'Possible health impact of paediatric scanning—a molecular epidemiology study'. The workpackage is focused on improving direct estimation of cancer risk following low doses of ionizing radiation from CT scanning in childhood and adolescence and to study the role of factors such as age and genetic and epigenetic variants that may modify this risk.

Recent articles

Recent literature also includes two large historical cohort studies that used clinical data both for exposure and for outcome (Mathews *et al* 2013, Pearce *et al* 2012). There are a number of other studies also that used either patient data and projections models or solely mathematical models to estimate risks. According to Chen *et al* (2014), who did an extended systematic review of all literature focused on head and neck CT, the two large cohort studies seem to provide the strongest data so far. All other studies contained less directly applicable data as their results were based on calculated projections or mathematical modeling. Some of these studies are summarized below.

- Brenner *et al* (2001).

  This was the first major study warning of CT risks to children, based on the Japanese exposure. According to this study, estimated lifetime cancer mortality risks attributable to the radiation exposure from a CT in a 1-year-old child are 0.18% (abdominal) and 0.07% (head)—an order of magnitude higher than for adults—although those figures still represent a small increase in cancer mortality over the natural background rate. In the United States, of approximately 600 000 abdominal and head CT examinations annually performed in children under the age of 15 years, a rough estimate is that 500 of these individuals might ultimately die from cancer attributable to the CT radiation.

  In response to this study, the American College of Radiology issued a statement urging parents not to refuse needed CT scans, especially for potentially life-threatening conditions like head and spine injuries, pneumonia complications and chest infections. At the same time, the organization lists conditions for which CT should not be the first choice. An example is suspected appendicitis in children, for which the group recommends that US be used first, followed by CT only if the US is equivocal.

- Pearce *et al* (2012) (UK study).

  This is a retrospective cohort study in which patients without previous cancer diagnoses were included. The patients were first examined with CT in National Health Service (NHS) centers in England, Wales, or Scotland

(Great Britain) between 1985 and 2002, when they were younger than 22 years of age. It is the first retrospective cohort study on the subject with personal information collected between 1985 and 2008 from the RIS database. Because no description of CT protocols and machine settings is available in RIS for this period of time, dose reconstruction was based solely on survey data. The results of the study showed that use of CT scans in children of about 50 mGy might almost triple the risk of leukemia and that doses of about 60 mGy might triple the risk of brain cancer. The cumulative absolute risks are small: for 10 years after the first scan for patients younger than 10 years, the estimated risk is one excess case of leukemia and one excess case of brain tumour per 10 000 head CT scans.

- Mathews *et al* (2013) (Australian study).

  The purpose of the particular study was to assess cancer risk in children and adolescents following exposure to low-dose ionizing radiation from diagnostic CT scans. The study was a population-based cohort data linkage study in Australia. The study included 10.9 million people, aged 0–19 years, who were identified from the Australian Medicare records. Overall cancer incidence was 24% greater for exposed than for unexposed people, after accounting for age, sex and year of birth. According to the study, the incidence rate ratio increased by 0.16 for each additional CT scan.
- Chen *et al* (2014).

  The authors performed a systematic review to evaluate the risk of malignancy associated with head and/or neck CT in infants, children and adolescents. They concluded that there is an impact of head and neck CT scan on subsequent risk of malignancy. Of course, as they commented, data regarding the magnitude of the effect of otolarygological imaging (i.e., temporal bone, sinus, neck) are limited.
- Meulepas *et al* (2018). The authors evaluated leukemia and brain tumor risk following exposure to low-dose ionizing radiation from CT scans in childhood by a nationwide retrospective cohort study of 168 394 children who received one or more CT scans in a Dutch hospital between 1979 and 2012. They found evidence that CT-related radiation exposure increases brain tumor risk. No association was observed for leukemia. Compared with the general population, incidence of brain tumors was higher in the cohort of children with CT scans, requiring cautious interpretation of the findings.

References

Brenner D J *et al* 2001 Estimated risk of radiation-induced fatal cancer for pediatric CT *Am. J. Roentgenol.* **176** 289–96

Brent R L 1980 Radiation teratogenesis *Teratology* **2** 281–98

Chen J X, Kachniarz B, Gilani S and Shin J J 2014 Risk of malignancy associated with head and neck CT in children: a systematic review *Otolaryngol. Head Neck Surg.* **151** 554–66

Chernobyl's Legacy: Health, Environmental and Socio-economic Impacts and Recommendations to the Governments of Belarus, the Russian Federation and Ukraine The Chernobyl Forum:

2003–2005 Second revised version. IAEA Division of Public Information, IAEA/PI/A.87 Rev.2/06-09181, April 2006

Donnelly E H, Smith J M, Farfan E B and Ozcan I 2011 Prenatal radiation exposure: background material for counseling pregnant patients following exposure to radiation *Disaster Med Public Health Preparedness* **5** 62–8

ICRP 2000 Pregnancy and medical radiation. ICRP publication 84 *Ann. ICRP* **30** 1–45

ICRP 2002 Basic anatomical and physiological data for use in radiological protection reference values. ICRP publication 89 *Ann. ICRP* **32** 1–265

ICRP 2003 Biological effects after prenatal irradiation (embryo and fetus). ICRP publication 90 *Ann. ICRP* **33** 1–200

ICRP 2007 The 2007 recommendations of the international commission on radiological protection. ICRP publication 103 *Ann. ICRP* **37** 2–4

Kippley J and Kippley S 1996 *The Art of Natural Family Planning* 4th edn (Cincinnati, OH: The Couple to Couple League), p 92

Lenton E A, Landgren B M and Sexton L *et al* 1984 Normal variation in the length of the follicular phase of the menstrual cycle: effect of chronological age *Br. J. Obstet. Gynaecol.* **91** 681–4

Mathews J D *et al* 2013 Cancer risk in 680000 people exposed to computed tomography scans in childhood or adolescence: data linkage study of 11 million Australians *Brit. Med. J.* **21** f2360

Mihm M, Gangooly S and Muttukrishna S 2011 The normal menstrual cycle in women *Anim. Reprod. Sci.* **124** 229–36

McCollough C H, Schueler B A and Atwell T D *et al* 2007 Radiation exposure and pregnancy: when should we be concerned? *Radiographics* **27** 909–17

Meulepas J M *et al* 2018 Radiation exposure from pediatric CT scans and subsequent cancer risk in the Netherlands *J. Natl. Cancer Inst.* [Epub ahead of print]

Mole R H 1993 The biology and radiobiology of in utero development in relation to radiation protection *BJR* **66** 1095–102

Muirhead C and Kneale G 1989 Prenatal irradiation and childhood cancer *J. Radiol. Prot.* **9** 209–12

Murki S and Sharma D 2014 Intrauterine growth retardation-a review article *J. Neonatal. Biol.* **3** 135

National Research Council 2006 *Health Risks from Exposure to Low Levels of Ionizing Radiation: BEIR VII Phase 2* (Washington, DC: The National Academies Press)

Noone A M, Howlader N and Krapcho M *et al* (ed.) 2018 *SEER Cancer Statistics Review, 1975-2015* (Bethesda, MD: National Cancer Institute) https://seer.cancer.gov/csr/1975_2015/, based on November 2017 SEER data submission, posted to the SEER web site, April 2018

Otake M and Schull W J 1998 Radiation-related brain damage and growth retardation among the prenatally exposed atomic bomb survivors *Int. J. Radiat. Biol.* **74** 159–71

Pearce M S *et al* 2012 Radiation exposure from CT scans in childhood and subsequent risk of leukaemia and brain tumours: A retrospective cohort study *Lancet* **380** 499–505

Public Health England 2017 National congenital anomaly and rare disease registration service: Congenital anomaly statistics 2015. Published by Public Health England

Public Health England 2018 Childhood Cancer Statistics: England Annual report 2018. Published by Public Health England

Reed B G and Carr B R 2015 The normal menstrual cycle and the control of ovulation [Updated 2015 May 22] *Endotext [Internet]* ed L J De Groot, G Chrousos and K Dungan *et al* (South Dartmouth (MA): MDText.com, Inc.), 2000

Stewart A, Webb J and Giles D *et al* 1956 Malignant disease in childhood and diagnostic irradiation in utero *Lancet* **2** 447–8

Stewart A, Webb J and Hewitt D 1958 A survey of childhood malignancies *Brit. Med. J.* **1** 1495–508

Stewart A M and Kneale G W 1970a Age-distribution of cancers caused by obstetric x-rays and their relevance to cancer latent periods *Lancet* **2** 4–8

Stewart A M and Kneale G W 1970b Radiation dose effects in relation to obstetric x-rays and childhood cancers *Lancet.* **1** 1185–8

Thierry-Chef I *et al* 2013 Assessing organ doses from paediatric CT scans-a novel approach for an epidemiology study (the EPI-CT study) *Int. J. Environ. Res. Public Health* **10** 717–28

United Nations Scientific Committee on the Effects of Atomic Radiation (UNSCEAR) 2013 Report to the General Assembly. Volume II: Scientific Annex B: Effects of radiation exposure of children

U.S. Department of Health and Human Services 2015 *Health Resources and Services Administration, Maternal and Child Health Bureau. Child Health USA 2014* (Rockville, Maryland: U.S. Department of Health and Human Services) Online at http://mchb.hrsa.gov/chusa14/

U.S. Department of Health and Human Services 2003 *Food and Drug Administration. Guidance for Industry and FDA Staff: Pediatric Expertise for Advisory Panels* (Rockville, MD: US Department of Health and Human Services, Food and Drug Administration, Center for Devices and Radiological Health)

Wagner L K, Lester R G and Saldana L R 1997 *Exposure of the Pregnant Patient to Diagnostic Radiations* 2nd edn (Medical Physics Publishing)

Ward E, DeSantis C and Robbins A *et al* 2014 Childhood and adolescent cancer statistics *CA: A Cancer J. Clinicians.* **64** 83–103

World Health Organization (WHO) 2016 Communicating radiation risks in paediatric imaging. Information to support healthcare discussions about benefit and risk WHO publication

Wilcox A J, Dunson D and Baird D D 2000 The timing of the 'fertile window' in the menstrual cycle: day specific estimates from a prospective study *Brit. Med. J.* **321** 1259–62

Chapter 3

Parameters that influence conceptus and paediatric patient radiation dose from radiodiagnostic procedures

Perisinakis Kostas and Papadakis Antonis

The radiation dose burden of an embryo/fetus or child associated with a radiodiagnostic procedure depends on many factors due to the complexity of modern radiodiagnostic systems and the high variation in size and anatomy of individuals during gestation and childhood. Deep knowledge and familiarization of the factors affecting radiodiagnostic exposures is prerequisite for the effective optimization of medical imaging procedures. All equipment-related and patient-related factors affecting a radiation dose burden of embryo/fetus or child from radiodiagnostic procedures are discussed in this chapter.

3.1 Radiography and fluoroscopy parameters that influence conceptus and paediatric dose

Thousands of pregnant patients are subjected to clinically justified radiography/fluoroscopy procedures each year and the trend is increasing. The average annual increase in the number of examinations has been reported as 6.8% for radiography and 10.6% for fluoroscopy, while radiography has been documented as the most frequent examination comprising 66% of the performed radiologic examinations (Damilakis *et al* 2001, 2003, Lazarus *et al* 2009). Radiography/fluoroscopy examinations are also frequent in paediatric patients. Children are more radiosensitive than adults. They have a longer life expectancy than adults and are therefore at a greater risk to the long-term harmful effects of radiation. Efforts should thus be given towards optimization of radiographic and fluoroscopic exposures to suppress conceptus and paediatric radiation burden to the minimum possible and minimize associated radiation risks for detrimental effects. In a new guidance, the U.S. Food

doi:10.1088/978-0-7503-1317-9ch3

and Drug Administration (FDA 2017) have recently recommended that medical X-ray imaging exams be optimized to use the lowest radiation dose needed. The factors that affect conceptus and paediatric radiation dose in radiographic and fluoroscopic exposures may be categorized in factors related to (i) the exposure parameters and the technology of the employed X-ray system, and (ii) the anatomical characteristics of the examined patients. Optimization of radiographic and fluoroscopic exposures requires deep knowledge and understanding of how all these factors affect dose. The experience and expertise of the radiologic technologist has a considerable influence on the way the examination is performed and on the dose given to the patient.

3.1.1 The tube potential (kVp)

The kVp selected at a given projection is under the control of the radiologic technologist. Patient dose is increased with kVp with a power of approximately 2.7, that is Dose $\propto$ kVp$^{\sim 2.7}$. Different kVp settings may be employed at different institutions for the same examination protocol. This suggests that the employed kVp is often not optimized. Several factors need to be considered in the selection of kVp for a particular examination protocol. For a constant detector exposure an increase in kVp should (i) decrease the tube current time product (mAs) required to achieve a constant detector exposure, (ii) decrease entrance skin dose and effective dose, (iii) decrease image contrast, and (iv) increase scattered radiation. As kVp is increased, contrast is reduced because the higher energy X-rays are less attenuated. The selection of kVp also depends on the tissue type and inherent contrast of the anatomical region being examined and to some extent on the pathology being investigated. Typically the ratio between the linear attenuation coefficient of cortical bone and muscle tissue is 4.2 at 40 keV and 2.1 at 80 keV, which yields a reduction of subject contrast by a factor of two (ICRP 23 1975).

The European guidelines (Council of the European Communities 1996) have influenced the setting of exposure parameters across the range of different radiographic views. In paediatric patients the use of a higher kVp is recommended for all radiographic exposures. For older tube-generator systems where short exposure times are not available, a slight lowering of kVp and use of additional filtration is recommended to achieve the required optical density. As stated in these guidelines, 'The soft part of the radiation spectrum that is completely absorbed in the patient, is useless for the production of the radiographic image and contributes unnecessarily to the patient dose'. There is evidence suggesting that a harder X-ray beam (higher kVp) can result in a lower effective dose for projections of the chest and the abdomen (Shrimpton 1988). However, as stated above, a harder beam may result in reduced inherent image contrast as fewer X-ray photons are attenuated (Shrimpton 1988, Martin 2007).

3.1.2 The tube load and exposure time product (mAs)

Patient dose from radiographic/fluoroscopic procedures is increased linearly with tube load (mAs in radiography or mA in fluoroscopy). With all other exposure parameters being the same, an increase in mAs should (i) increase detector exposure (equivalent air kerma) and exposure index (EI) as more X-ray photons will reach the detector, (ii) improve image quality through increased signal-to-noise (SNR) ratio and contrast-to-noise ratio (CNR), and (iii) increase patient radiation dose. To select the appropriate mAs setting for a given examination protocol, radiologic technologists need to consider that the optimum target detector exposure is achieved for the anatomical region of diagnostic interest. In manual exposures, mass-based mAs technique charts need to be carefully followed so that mAs is tailored to the each patient's body habitus.

3.1.3 The automatic exposure control (AEC) system

The AEC in digital radiology systems operates identically to that used in screen-film systems. It is important that the AEC is calibrated to match the digital detector system before use. AEC systems use three or more ionization chambers that are preprogrammed on phantoms and aim to automatically determine the exposure time. AEC systems control the total mAs of the examination, but the radiologic technologists are still in charge of selecting the optimum mA and kVp for an examination. Although AEC activation is recommended in most radiographic acquisitions there are times the system should not be used. For instance, when the anatomy of interest is smaller than the active area of the AEC's detector cell, AEC will not operate effectively. In such a case, the areas of the detector not covered by the patient's body receive more radiation than the anatomy to be imaged, causing early termination of the exposure, which results in images with a high quantum noise. This is particularly important in paediatric radiography. When AEC detector cells are within an edge of the patient's body, such as the clavicle for instance, early termination of the exposure is provoked. The result is again insufficient exposure to the digital detector causing images with a high quantum noise. Further, the presence of image artifacts, caused for instance by metallic orthopedic hardware, can contraindicate the use of AEC. They need to be moved away from the imaged area, otherwise they generate unexposed areas over the AEC detector cells causing an increase in exposure time and consequently overexposure to the patient. It is apparent that the use of AEC requires accurate patient positioning and periodic calibration of the AEC. Radiologic technologists should ensure that the anatomy to be imaged covers the AEC's detector cells, and put emphasis on accurate patient positioning before exposure. It is important that radiologic technologists follow the department protocols and exposure technique charts regarding the proper use of AEC.

3.1.4 Selection of tube-to-patient and tube-to-detector distance

The X-ray beam intensity decreases with distance from the tube according to inverse square law. The free in air dose at the position of the patient's surface is higher than the dose at the position of the detector by a factor of [(*focus to detector distance*)/(*focus to surface distance*)]^2, that is (FDD/FSD)^2. As this ratio is decreased by

Table 3.1. The effect of FDD on S_D, entrance area and DAP.

FDD (cm)	FSD (cm)	mAs	S_D (mGy)	Entrance area (cm × cm)	DAP (mGy·cm^2)
50	25	25	20.4	21.5 × 21.5	9245
75	50	56	11.5	28.6 × 28.6	9245
100	**75**	**100**	**9.1**	**32.5 × 32.5**	**9245**
150	125	225	7.4	35.8 × 35.8	9245
200	175	400	6.4	37.6 × 37.6	9245

moving the patient further away from the tube, the surface to detector dose ratio (S_D/D_D) is reduced. The S_D/D_D ratio is also kept small by ensuring that the patient is as close to the detector as possible. Table 3.1 demonstrates the effect of FDD on S_D for a patient with an abdominal anterior-posterior thickness of 25 cm. Calculations were performed using the inverse square law by considering a S_D of 9.1 mGy at a FDD of 1 m and a constant field size on the detector (43 cm × 43 cm). The dose at the detector in the absence of the patient would be 5 mGy. The dose area product at the detector would be 43 × 43 × 5 = 9245 mGy·cm^2.

As demonstrated in table 3.1, DAP is constant regardless the FDD. This is due to the fact that, S_D is reduced with the square root of FSD, while beam area increases with the square of FDD. Given that the effective dose may be considered proportional to DAP, changes in FDD do not have a strong effect on patient dose. However, S_D values suggest that an increase in FDD and FSD result in significant reduction of skin dose.

3.1.5 Setting up the collimation and field size

Ideally the X-ray beam should be limited to the area of clinical interest. Accurate collimation reduces the dose to the patient and improves image quality since internal scatter is reduced. Collimation can have a major effect on doses to conceptus and radiosensitive organs of paediatric patients, especially when these organs are located adjacent to the examined anatomical region. Large field sizes increase scatter and consequently patient dose. Reduction of the field size may require a small increase in the mAs to reach the same detector exposure because of the reduced scatter. Conceptus dose from an abdominal radiographic or fluoroscopic procedure performed on the expectant mother is considerably increased with beam field size. This is owing to the increased internal scatter produced by the larger field sizes. It should be noted that a large field size impairs image quality due also to the increased internal scatter. In addition, the potential that conceptus lies closer to or partially within the primary exposed region, is higher with a large compared to a small or well collimated field size. The field size should thus be well collimated to the anatomical region of diagnostic interest so that conceptus dose is reduced and image quality is enhanced.

3.1.6 Projection

The dose to conceptus and radiosensitive organs of paediatric patients can vary significantly at various projections. Conceptus dose in a posterior anterior (PA) radiograph is substantially lower than in anterior posterior (AP) radiograph since spine and pelvic bones attenuate a significant portion of the X-ray beam before it reaches conceptus. It should be noted that the above ignores any difficulties in performing an abdominal radiograph on a pregnant patient in the PA projection. Moreover, breast equivalent dose in female chest radiography is higher for the AP than the PA.

3.1.7 Beam quality

kVp and tube filtration (inherent + added filtration) determine the mean energy of the emitted photons, which in turn determine the penetrability of the X-ray beam. A minimum of 2.5 mm Al equivalent total filtration is required by regulations to eliminate low energy photons from the X-ray beam and ensure low skin dose. It should be noted that the added filtration reduces patient dose but decreases image contrast. Many modern fluoroscopic systems have the ability to automatically adapt total filtration by using additional filters (commonly 0.1 mm–0.9 mm of Cu) to maintain constant photon flux on the detector based on the patient's body habitus and projection angle (Brosi *et al* 2011). The goal is to generate images of acceptable contrast at the highest possible kVp.

Evidence that filtration reduces the dose to the patient undergoing an abdominal radiographic exposure is demonstrated in table 3.2, which lists the surface dose (S_D) values on the anthropomorphic adult phantom, 16 cm thickness, at a constant dose on the detector. This phantom may simulate the abdomen of a thin female and 8 cm thickness may approximate the position of the ovary or conceptus. In table 3.2, the dose values at 8 cm (*D* at 8 cm) are relative to the 0.5 mm Al additional filtration. The S_D can be reduced by approximately 50% (31.8–16.5 μGy) by the addition of ~3 mm Al at 75 kVp, while dose at 8 cm is only reduced by 12% (4.7–4.1 μGy). This dose saving is achieved by the first 3 mm of added filtration, while the addition of further filtration would have less effect. The amount of additional filtration added is also limited by the heat loading capacity of the tube. Filtration attenuates also higher energy photons. For a small filtration increase no alteration of mAs is required, however, if too much filtration is added the mAs values required would reach the tube's heat capacity. Another limiting factor in the addition of a too thick

Table 3.2. The effect of additional filtration on the surface dose relative to the dose at 8 cm depth.

kVp	Additional filtration (mm Al)	S_D (μGy μGy^{-1} of exit dose)	Relative S_D (%)	*D* at 8 cm (μGy μGy^{-1} of exit dose)
75	0.5	31.8	100	4.7
75	1.5	23.5	76	4.4
75	3.5	16.5	54	4.1

filtration is the image contrast. The lower energy components of the beam are more selectively attenuated by different tissues. If too many of these components are removed by filtration then image contrast may degrade. Nevertheless, the degradation of image contrast for a 2–4 mm Al is virtually imperceptible. Thus, for an X-ray machine that already has a 2.5 mm Al total filtration, the employment of 2–3 mm of additional filtration will decrease patient S_D with only a minor effect on image contrast.

3.1.8 Anti-scatter grid

When scattered radiation is considered significant (that is in thick body anatomies), then for a constant detector exposure, the use of an anti-scatter grid will (i) improve image contrast by decreasing scattered radiation reaching the detector, and (ii) increase radiation dose to the patient (ICRP 103 2007). The higher sensitivity in digital imaging systems to low levels of radiation exposure makes the use of anti-scatter grids more important to ensure high quality images in DR. However, the conditions under which a scatter reduction grid should be employed are not universally accepted. A widely quoted source suggests that a grid should be used whenever the body part is 10 cm or greater in thickness and when the tube potential is higher than or equal to 60 or 70 kVp (Carlton and Adler 1996). This dimension corresponds roughly to the thorax thickness of a 6-month-old infant (Beck and Rosenstein 1979). It is a common practice, however, to routinely image older infants on the table top or bedside without a grid at an acceptable image quality. The use of a grid is discouraged when the imaged patient's body anatomy is less than 10–12 cm thick. The proportion of scattered radiation in such cases is so low that no discernible improvement in resolution can be achieved with a grid. In addition, it should be stressed that the use of anti-scatter grids results in an increase of dose to the exposed patient tissues by a factor of 2–5. The use of scattered-radiation grids is not thus recommended with paediatric patients and they should be used only when the body part thickness dictates their need. The use of a grid with a wrong line rate or improper orientation can result in image artifacts such as interference patterns with the digital sampling rate or with the display rate. For the same grid ratio, grids with a high line rate, enough to avoid being resolved by the digital sampling, are expensive and impose a greater dose burden than grids with a lower line rate. A parallel grid delivers higher doses than a focused grid.

The air gap technique may be used as an alternative to the scatter reduction grids. The image detector is moved away from the patient so that less scatter radiation is recorded by the detector. The tube-to-detector is increased so that there is little additional geometric magnification. With the air gap technique there is no real dose saving for the patient, however, the images are produced without scatter or grid lines.

3.1.9 Detector technology

Another factor that affects conceptus and paediatric radiation dose is the efficiency of the detector employed. The detector efficiency is determined by the speed of the screen-film combination in conventional radiography, the image intensifier in conventional fluoroscopy and the detector technology in digital radiography/ fluoroscopy.

3.1.10 Fluoroscopy

Fluoroscopic procedures result in higher doses compared to radiography. The two principal factors that affect conceptus and paediatric patient dose in fluoroscopy are X-ray dose rate and beam-on time. The radiation dose is proportional to both these factors, while deterministic effects have been documented in fluoroscopic procedures that high dose rates have been combined with prolonged fluoroscopy times (ICRP 85 2001).

3.1.11 Dose rate

Patient dose rates vary on the detector technology, age, quality of the X-ray unit, the machine setting established by the manufacturer and the exposure factors selected by the radiologic technologist. The most important factor is the dose rate recorded by the detector. A cesium-iodide based image intensifier detector system operates adequately at an input dose rate of 10 to 30 μGy min^{-1}. This dose rate increases with the age of the image intensifier detector as the X-ray to light photons conversion factor decreases. As this conversion factor decreases, the input dose rate has to be increased to maintain adequate image brightness. With automatic brightness control systems the image intensifier dose rate is mostly dependent on the setting established by the manufacturer. The system automatically tunes the kVp and/or mA to reach the preset input dose rate and consequently a preset image brightness regardless of patient body habitus.

3.1.12 The fluoroscopy beam-on time

The magnitude of conceptus and paediatric radiation dose in fluoroscopic procedures is principally determined by the beam-on time (fluoroscopy time). The beam-on time along with the kinetic energy released in matter (KERMA) rate (mGy min^{-1}) of the machine's output determine the KERMA (mGy) during the examination. The KERMA value is multiplied by the field size employed during the examination to estimate the total dose area product (mGy · cm). The two major factors that affect the beam-on time are the skill and experience of the radiologist, and the complexity of the examination. Other equipment-related factors may also contribute to beam-on time. Samara *et al* have reported that therapeutic ERCP in pregnancy may require 2.1 to 6.9 min of fluoroscopy time (Samara *et al* 2009). Theocharopoulos *et al*, in a study on conceptus dose during fluoroscopically guided surgical treatments of spinal disorders, have reported that 4 min and 11 min of fluoroscopy time are required for pedicle screw and kyphoplasty operations, respectively (Theocharopoulos *et al* 2006). Radiologists should put effort into minimizing the duration of the imaging time, without compromising the benefit of the examination. The X-ray unit itself has a minor influence on beam-on time, provided that image quality is sufficient for the examination performed. The use of last image hold and/or single shot X-ray modes display the last fluoroscopy frame, which may be retained on the screen as a reference. Alternatively, a number of frames may be averaged during a short pulse to generate an image that is updated either automatically or each time the radiation pulse is repeated. Conceptus dose from a fluoroscopic procedure performed on the expectant mother may be significantly reduced when a low pulse rate fluoroscopy mode is selected instead of continuous fluoroscopy. These acquisition techniques result in substantial dose reduction. A small loss in image quality compared to the live fluoroscopy mode may, however, be noted. The exposure settings for cine acquisition have a significant influence on patient dose. In cardiac catheterization procedures of paediatric patients, cine acquisition has been documented up to 59% of the exposure time.

3.1.13 Magnification

A significant factor, which also affects conceptus dose and paediatric radiation dose, is the image magnification. Image magnification is directly related to the distance between the patient and the detector. The magnification factor is kept at the minimum when the patient is located closest to the detector. As an estimate of how magnification affects dose, consider that when the patient-to-detector distance is doubled, conceptus dose is increased by a factor of 1.8, while when the patient-to-detector distance is tripled, conceptus dose is quadrupled.

3.1.14 Automatic exposure control

Modern fluoroscopic systems enable automatic adaptation of kVp, mA, optical aperture (image intensifiers), pulse width (PW), beam filtration and the input dose to the detector. The automatic control of all the above parameters makes modern fluoroscopy systems complex. It is essential that clinical medical physicists thoroughly understand the operational characteristics of modern AEC systems in the attempt to optimize radiation dose and image quality in fluoroscopic procedures.

3.1.15 The entrance exposure rate to the detector (EERD)

For both image intensifier and flat panel based fluoroscopic systems the energy distribution of the photons incident on the detector and the detective quantum efficiency of the detector determine the photon fluence that will be recorded at the input of the detector. These factors determine the EERD. EERD is not subject to a regulation. A survey conducted by the American Association of Physicists in Medicine, task group 11 for fluoroscopy systems in clinical use, revealed that the EERD varies in a wide range from 0.18 μGy s^{-1} to 9.1 μGy s^{-1} (Boone *et al* 1993). For image intensifier based continuous fluoroscopy at 30 fps with a 30 cm diameter FOV, an acceptable EERD value should be 265 nGy s^{-1}. Flat panel based fluoroscopy systems need 2–4 times the EERD required by conventional image intensifier systems due to the small size of the pixels and the increased noise arising from the read-out array electronics (Yadava *et al* 2008, Cowen *et al* 2008). To minimize elevated EERD, flat panel based systems should incorporate increased added filtration. However, in single frame or serial acquisition imaging modes, which employ a much higher input dose per frame compared to fluoroscopy mode, the flat panel detector based systems can operate at a lower EERD than that of an image intensifier.

3.1.16 The patient skin entrance exposure rate (SEER)

Patient skin entrance exposure rate depends on the X-ray generator operating control curve, which defines how kVp is modified on the varying attenuation produced by patients at varying body habitus, different projection geometries and different anatomical regions. Patient SEER depends also on beam filtration, the tube-to-patient distance, the table attenuation, the grid bucky factor and examined anatomical region. A single fluoroscopic system can have many operating control curves based on the examination protocol, the examined anatomical region and the required image quality. Figure 3.1 demonstrates representative X-ray generator control curves of a typical flat panel based fluoroscopy angiographic system.

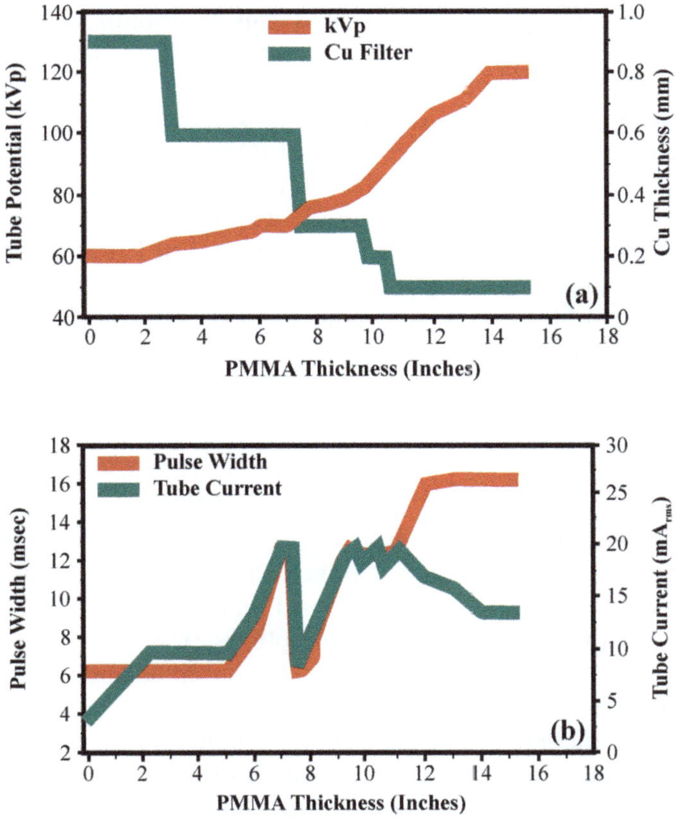

Figure 3.1. X-ray generator control curves for a typical flat panel based digital fluoroscopy angiographic system. Plots show how kVp and Cu thickness filtration (a) and PW and mA (b) vary on attenuation. Adapted from AAPM Task Group 125 (2012).

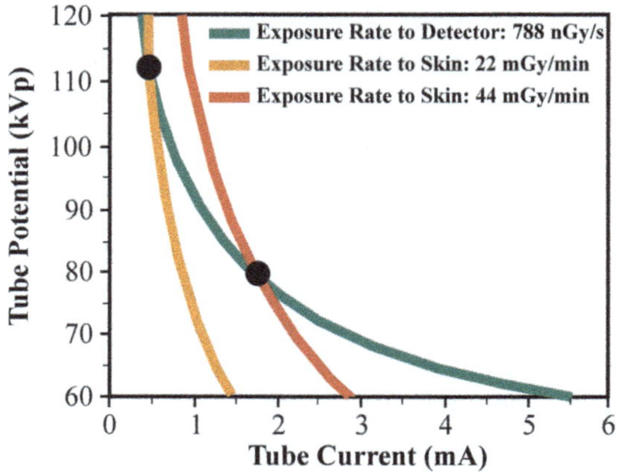

Figure 3.2. kVp versus mA required to maintain a constant EERD at 0.78 μGy s^{-1}, and a constant SEER at 22 mGy min^{-1}, and at 44 mGy min^{-1}, using a 20 cm thick water phantom. The two crossing points indicate the kVp, mA values that provide the same EERD but with patient doses differing by a factor of two. Adapted from Gagne and Quinn (1995).

Zamenhof demonstrated that image SNR can be optimized if an appropriate combination of kVp and beam filtration is achieved (Zamenhof 1982). Omnasch *et al* have demonstrated that image CNR was independent on beam filtration. Thus, thicker filters could be incorporated in the X-ray beam to reduce SEER, while lower kVp could be used to improve image contrast (Omnasch *et al* 2004). It should be noted that SEER does not depend on EERD. Figure 3.2 shows the kVp, mA values that would provide a EERD at 788 nGy s^{-1} with a 20 cm water phantom as the attenuator. Further, the kVp, mA values that provide a SEER of 22 mGy min^{-1} and 44 mGy min^{-1} are shown. Although the dose delivered to the detector is identical at the kVp, mA values of the two crossing points, the SEER value differs by a factor of two. It is important that medical physicists have knowledge and understanding of all parameters that affect patient dose for the fluoroscopic system they use since the effect of all these factors on EERD and SEER is not easily realized.

3.1.17 Location of conceptus relative to the X-ray beam

The magnitude of conceptus dose from radiographic or fluoroscopic procedures performed on the expectant mother is principally determined by the inclusion or not of the conceptus in the primarily exposed anatomical region. If the uterus is positioned outside the field of view, the conceptus is exposed to scattered radiation only and the conceptus dose is minimal. Higher conceptus dose values occur when the uterus is positioned within the field of view. In this case, the radiation dose to a conceptus from a radiographic or fluoroscopic examination depends on the patient's body habitus, the projection angle (anterior-posterior, posterior-anterior, or lateral), the depth of the conceptus from the skin surface and exposure factors. In abdominal radiographic or fluoroscopic procedures, that conceptus lies completely or partially within the exposed area, conceptus dose is much higher compared to thorax examinations. The dose to the conceptus for a specific examination or projection may vary by a factor of ten. Extra-abdominal radiographic/fluoroscopic procedures result in a conceptus dose lower than 1 mGy, while abdominal radiographic procedures may result in conceptus doses of 1 mGy to 4 mGy. Studies have documented that abdominal fluoroscopically guided procedures performed on a pregnant patient may result in the conceptus dose occasionally exceeding 10 mGy.

3.1.18 Anatomic characteristics of the pregnant patient that affects conceptus dose

Conceptus dose from a radiographic or fluoroscopic procedure performed on an expectant mother is strongly affected by her body size, since for large patients higher exposure parameters are required to obtain the desired image quality. Special care must be taken in overweight females at early pregnancy and average or overweight females in late pregnancy. Charts with exposure parameters that prescribe the kVp, mAs settings that are required to generate images of diagnostic quality at a low conceptus dose need to be employed to avoid repetition of exposures. To correctly apply these exposure charts, radiographers need to be well trained to determine a patient's body size by accurately measuring abdominal circumference.

Conceptus dose depends also on the size of conceptus and the depth below the anterior abdominal surface that the conceptus is located. The dose to the uterus can be taken to be the dose to the whole body of the conceptus and is thus, taken to be equivalent to the effective dose to the conceptus. Perisinakis *et al* have shown that uterus location differs from conceptus location during organogenesis and the early fetal period by 1 cm (Perisinakis *et al* 1999). The size of the conceptus is mainly determined by the gestational stage at the time of exposure. In late pregnancy that conceptus has grown up and has reached its maximum size, it is more likely that it will be included in the primary exposed volume. Radiographers should attempt to avoid exposure of the conceptus by collimating the X-ray beam to the anatomical region of interest. Conceptus depth varies considerably during early pregnancy depending on body habitus and anatomical characteristics of the expectant mother and the degree of bladder fullness. The higher the distance between the anterior abdominal surface of the expectant mother and conceptus, the lower the conceptus dose in an AP X-ray beam projection. Correspondingly, the higher the distance between the anterior abdominal surface of the expectant mother and conceptus, the higher the conceptus dose in a PA X-ray beam projection. It is not straightforward to accurately determine conceptus depth. Ragozzino *et al* have assumed that in early pregnancy conceptus depth may be considered equivalent to uterus depth, which is approximately 12 cm deep from the anterior abdominal surface (Ragozzino *et al* 1981). Stovall *et al* proposed that conceptus depth should be 10 cm from the anterior abdominal surface (Stovall *et al* 1995). If it is taken into account that attenuation of 4 cm of soft tissue alters the measured dose by a factor of two then it is obvious that to accurately assess conceptus dose, correct measurement of conceptus depth is needed (Harrison 1981). Conceptus depth varies considerably also on the degree of bladder fullness. Perisinakis *et al* have demonstrated that uterus and conceptus depth are strongly dependent on bladder status (Perisinakis *et al* 1999). This dependence is more pronounced during organogenesis and the early fetal period.

In particular, when the bladder is empty, conceptus depth from the anterior abdominal surface ranges from 4 to 8 cm. In contrast, when the bladder is full, conceptus depth ranges from 6 to 10 cm. This shows that assuming that uterus dose equals the conceptus dose should be therefore considered a rather rough approximation.

Table 3.3 tabulates typical conceptus doses following radiographic/fluoroscopic procedures at each trimester of gestation.

3.1.19 The effect of paediatric patient's size

Body habitus and tissue composition result in differences in X-ray beam attenuation. Muscle tissue is denser than fat tissue, and requires an increase in exposure parameters so that the beam can adequately penetrate the muscle tissue, regardless of a patient's size. Simple reconfiguring the examination protocols applied in adults for use in children is not efficient. The somatometrical characteristics of children's anatomies vary much more than those of adults. This makes it difficult to determine the appropriate exposure technique because patient size depends not only on a child's age but also on the child's individual characteristics. In addition to the variation in the body size from one child to another at the same age, children's body parts grow at different rates. For instance, an infant's femur is one-fifth the size of that of an adult. Whereas an infant's skull grows slowly up to triple in size by adulthood.

Table 3.4 tabulates typical entrance surface dose, DAP and effective doses in paediatric patients following typical radiographic/fluoroscopic procedures.

3.2 CT parameters that influence conceptus and paediatric dose

The use of CT imaging has been steadily increased over the last few decades following the rapid evolution of multidetector CT technology (Kocher 2011). The same increasing trend in CT utilization has been observed for paediatric (Pola 2018) and pregnant patients (Lazarus 2009, Goldberg-Stein 2011). The radiation exposure associated with CT imaging is relatively high compared to other common X-ray diagnostic procedures. Currently, CT is the main contributor to exposure of the public from medical radiation exposures (Berrington de Gonzalez 2009). Reduction of radiation dose from CT and optimization of CT examinations has been, and still is, a hot topic for the radiology and medical physics community.

CT exposure is rather complex compared to common radiographic exposures. In addition, modern CT scanners are equipped with advanced complex dose sparing features/techniques that may significantly affect patient dose (Kubo 2019). All these exposure factors/features and dose sparing tools/techniques may considerably affect radiation dose to the examined patient during a CT examination as well as the image quality of the resulting image series. Since all these factors may not act synergistically, optimization of a CT examination is a particularly difficult task. The factors affecting radiation dose burden to an exposed child or conceptus from a CT exposure are essentially the same factors affecting patient dose from CT and may be categorized in:

Table 3.3. Typical conceptus doses following radiographic/fluoroscopic procedures at each trimester of gestation.

		Conceptus dose (mGy × 10^{-3})		
Examination	Projection	First trimester	Second trimester	Third trimester
Chest radiography (Damilakis *et al* 2003, Tirada *et al* 2015)	AP	2.8	23	36
Head, neck radiography (Tirada *et al* 2015)	Cervical spine AP and LAT		<1	
Extremities radiography (Tirada *et al* 2015)	Upper extremity 2 projections		<1	
	Lower extremities 2 projections			
Fluoroscopically assisted surgical treatment of hip fractures (McCollough *et al* 2007)	PA	42	—	—
Abdomen radiography (McCollough *et al* 2007) (33 cm patient thickness)	AP	3000	—	—
Lumbar spine radiography (McCollough *et al* 2007)	AP and LAT	1000	1000	>1000
Intravenous pyelography (McCollough *et al* 2007) (21 cm patient thickness)	4 AP projections		6000	
Double-contrast barium enema study (McCollough *et al* 2007)	4 min fluoroscopy + 12 digital spot images		7000	
Cardiac catheter ablation (Damilakis *et al* 2003)	0.58 min GHPA 23 min PA 5.3 min RAO 10.2 min LAO	204	300	557
Fluoroscopically guided kyphoplasty (Theocharopoulos *et al* 2006)	10.1 min AP and LAT at 5th lumbar vertebrae	28 800	31 800	104 000
Therapeutic ERCP	2.1–6.9 min fluoroscopy	33 600	7450	—

Table 3.4. Entrance surface dose, DAP and effective doses in paediatric patients following typical radiographic/fluoroscopic procedures.

Examination	Procedure type	Age group	Entrance surface dose (mGy)	DAP ($Gy \cdot cm^2$)	Effective dose (mSv)
Chest radiography (Yakoumakis *et al* 2007)	PA projection	0.5–2 years	0.263	—	0.049
		3–7 years	0.410	—	0.073
Chest radiography (Freitas *et al* 2004)	AP projection	0.3–12 years	0.12	—	—
Pelvis radiography (Yakoumakis *et al* 2007)	AP projection	0.5–2 years	0.358	—	0.047
		3–7 years	0.669	—	0.094
Cardiac catheterization (Dragusin *et al* 2008)	Therapeutic	0–30 days	—	4.8	17.2
Cardiac catheterization (Bacher *et al* 2005)	Arterial septal defect	2 years	34.2	4.63	6
Cardiac catheterization (Yakoumakis *et al* 2013)	Arterial septal defect	8.5 years	91.0	40.3	40

Exposure parameters;
CT equipment characteristics and dose sparing tools/techniques;
Patient-related factors.

3.2.1 Exposure parameters affecting radiation dose to conceptus or children exposed during a CT examination

Exposure parameters that affect the dose to a conceptus or child exposed during a CT examination are commonly prescribed in the standardized preset CT imaging acquisition protocol but the operator of the CT equipment may alter them prior to the examination.

Tube load

Tube load (in mAs) is defined as the product of the tube current of the X-ray tube (mA) and the rotation time (s). Tube load defines the number of photons emitted per rotation, i.e. the photon flux of the CT beam. The higher the tube load, the higher the number of photons passing through patient tissues during CT exposure. The dose absorbed by patient tissues is therefore directly proportional to tube load if all other parameters are held constant (Maldjian and Goldman 2013). Patient radiation dose may be reduced by decreasing the mAs value, which may be accomplished by either decreasing the tube current (mA) or decreasing rotation time (i.e. increasing rotation speed). However, the tube load is associated with the noise of the resulting CT images. Noise is proportional to $\frac{1}{\sqrt{\mathrm{mAs}}}$ which means that if mAs (i.e. patient dose) is reduced by 1/2, the noise will be increased by about 40%. Therefore, operators should be very cautious when considering mAs reduction to avoid nondiagnostic noisy images. This issue of mAs optimization has been addressed by the introduction of automatic mA modulation techniques discussed later in this chapter.

Tube voltage
Tube voltage (in kV) is the constant voltage applied in the X-ray tube to accelerate electrons hitting the anode and affects the quality and intensity of the X-ray beam produced, i.e. the mean photon energy the and number of photons making up the CT beam. Most CT scanners allow acquisition at 80, 100, 120 or 140 kV, with the most frequently used value being 120 kV. The higher the kV setting, the higher the mean energy of produced photons and also, the higher is the X-ray tube output, i.e. the number of photons produced. Therefore, kV setting affects both the penetration ability of the produced CT beam and the photon flux of the CT beam. The use of high kV deteriorates CT image contrast but is necessary in the case of large bodies that require a CT beam of high penetration power to achieve acceptable noise at relatively low dose. The dose absorbed by patient tissues is proportional to kV raised to an exponential power that ranges from 2.5 to 3.1 depending on the patient's body size (Maldjian and Goldman 2013). Decreasing tube voltage from 120 kV to 80 kV and keeping all other exposure parameters constant results in a patient dose reduction of the order of 65%. However, decreasing kV results in a concomitant increase in noise, which is commonly counterbalanced by increasing the mAs setting. Given that the use of a CT beam of lower mean photon energy may increase imaging contrast between different tissues, the mAs increase required to maintain the same contrast-to-noise ratio may result in an overall lower dose for the examined patient compared to that when using a high kV technique. However, the appropriateness of employing low-kV depends on the size of the patient, the use of a contrast medium and the specific clinical diagnostic task. The low-kV technique should be applied with caution. In general, the use of 'low kV–high mAs' technique may be beneficial regarding patient dose for small and average size patients as well as, in contrast-enhanced CT examinations. Some modern scanners are equipped with automatic kV selection tools discussed later in this chapter.

Beam collimation
Modern multi-slice CT scanners allow acquisition using beams of different beam collimation. Beam collimation refers to the number of active detector elements times the width of each element at the isocenter (also called detector collimation or effective detector row thickness). For example, assume a scanner equipped with 64 detector rows along the z-axis with each detector row receiving a fraction of the CT beam that is 0.625 mm wide at the isocenter. That means that the maximum selectable beam width at the isocenter is $64 \times 0.625 = 40$ mm. Such a CT system, however, allows the operator to set the beam collimation to 32×0.625 mm, which is 20 mm wide at the isocenter. Given that the collimated X-ray beam has a profile that is of gaussian shape and not rectangular, the actual beam width at the isocenter is always higher (by few mm) than the total width of active detectors to maintain approximately the same X-ray beam intensity falling on the edge detector rows as on

the middle detector rows. Consequently, a small fraction of the CT beam, commonly referred to as penumbra, falls outside the edges of the active detector array. The penumbra of the CT beam goes undetected and therefore does not contribute to the generation of CT images but only increases the radiation dose to patient tissues (Perisinakis 2009). This is referred to as over-beaming in multi-slice CT imaging. The fraction of the beam falling on the active detector array is referred to as geometric efficiency. The extra beam width required to exclude penumbra from falling on edge detector rows is essentially the same regardless of the total beam collimation. Therefore, wide CT beams are associated with superior geometric efficiency and therefore a slightly lower dose for the patient if all other parameters are held constant. Namely, the 64 × 0.625 mm beam collimation is more dose efficient than 32 × 0.625 mm. Also, for a certain beam collimation the use of small focus during acquisition is associated with less penumbra and therefore higher geometric efficiency and less radiation burden for the patient (Perisinakis *et al* 2009). In most systems the size of the focal spot is automatically selected depending on the mAs setting. For mAs values above a threshold the large focal spot is employed.

In the previous CT system, a beam collimation of 32 × 1.25 mm is also selectable that is of the same width at the isocenter as 64 × 0.625 mm beam collimation and therefore there is no difference regarding dose due to over-beaming between these two collimations. The patient radiation dose does not directly depend on detector collimation. However, the selected detector collimation defines the minimum selectable reconstructed slice thickness, since the reconstructed slice thickness is commonly the product of the detector collimation and an integer $\geqslant 1$. Namely, if the operator selects 32 × 1.25 mm = 40 mm beam collimation instead of 64 × 0.625 mm = 40 mm with all other exposure parameters held constant, the patient dose remains the same but the minimum allowed reconstructed slice thickness is 1.25 mm instead of 0.625 mm, respectively. The minimum reconstructed slice thickness may indirectly affect the dose. Namely, images of 0.625 mm slice thickness may require a higher mAs value during acquisition compared to 1.25 mm thick images in order to keep noise at an acceptable level (Maldjian and Goldman 2013).

Pitch

Pitch in sequential step-and-shoot CT acquisition is defined as the distance traveled by the CT table along the patient's axis in-between successive sequential CT acquisitions divided by the beam collimation. Pitch in helical CT acquisition mode is defined as the ratio of the distance traveled along the patient's axis during the time for one full rotation divided by the collimated CT beam width at isocenter. For example, if the pitch is set to one and the beam collimation is set to 64 × 0.625 mm = 40 mm then:

In sequential acquisition: the table will travel 40 mm along the patient's axis in-between successive scans.

In helical acquisition: the table will travel 40 mm along the patient's axis within the time required for a full 360° rotation.

Increasing the pitch from 1 to 1.25 means that the table will travel 50 mm along the patient's axis in-between successive scans of a sequential acquisition or in one rotation time in helical acquisition. In either case, the same amount of photons will interact with a patient's body region that is 20% longer compared to the body region exposed when pitch = 1. Apparently, increasing pitch results in a lower number of photons passing through the patient's tissues due to the rapid movement of the table. Thus, the patient's dose is inversely proportional to pitch if all other exposure parameters are held constant. For example, increasing pitch from 1 to 1.25 results in a patient's dose reduction of 20% while decreasing pitch from 1 to 0.75 results in a patient's dose increase of 33%. It has to be stressed that, when pitch is increased while all other parameters remain constant, the quality of the produced images deteriorates since noise is increased and effective slice thickness is also increased. Therefore, for increased pitch CT acquisition, the mAs value is commonly elevated to maintain image quality

3.2.2 CT equipment characteristics/dose sparing affecting radiation dose to conceptus or children exposed during a CT examination

CT beam filtering

In all CT systems, the produced CT beam is filtered prior to passing through a patient's body. There are two types of filters used in combination: the flat filters and the beam shaping (or bow-tie) filters. Flat filters are used to suppress the low energy photons' component of the CT beam as shown in figure 3.3. Low energy photons that have very low possibility to penetrate a patient's body are thus removed and therefore patient tissues absorb a lower amount of dose. The width and combination of materials of flat filters affect the penetration power of the CT beam by altering the

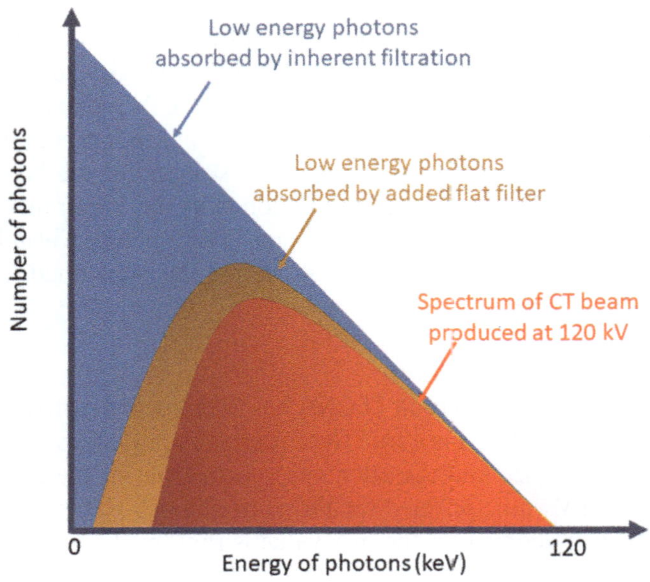

Figure 3.3. The effect of flat filtering on a CT beam spectrum.

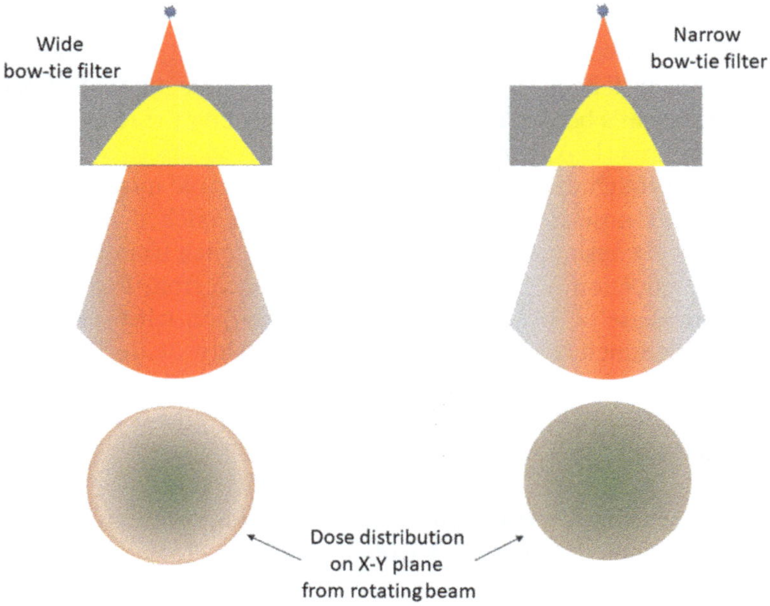

Figure 3.4. The effect of a bow-tie filter on dose distribution throughout the X–Y plane of the imaged body.

mean photon energy of the beam. CT vendors commonly provide more than one (usually two or three) selectable flat filters, each being a combination of different materials (usually Al, Cu, tin, Teflon, etc) of standard thicknesses. Each flat filter is combined with a bow-tie filter. The bow-tie filter (or wedge or beam shaping filter) aims to shape the CT fan beam at the X–Y plane to reduce dose at the periphery of the human body to be imaged, as illustrated in figure 3.4.

Some modern CT scanners provide three different bow-tie filters (e.g. large, medium and small) to match the size of the imaged body region in order to achieve a more or less uniform dose deposition throughout the X–Y plane (Perisinakis 2013). The appropriate flat filter + beam shaping filter combination is commonly preset on imaging protocols in association with the patient size and/or the body region to be imaged. Namely, a different filter combination may be recommended for adult body, cardiac or paediatric CT examinations. The operator usually selects the exposure acquisition protocol and does not change the predefined (recommended) filter combination.

Automatic tube current modulation

Automatic tube current modulation (TCM) has been incorporated to CT scanners more than 20 years now. TCM is a tool for automatic adaptation of tube current on the basis of the patient size and the attenuation properties of the body region to be imaged for a given kV value selected for the examination. Namely, when the TCM tool is applied during acquisition the mA value is rapidly changed during scanning and an increased or decreased mA value is employed when the CT beam penetrates body regions of high or low attenuation ability, respectively. In fixed current CT acquisition protocols all attenuation data are collected with a beam of the same intensity irrespective of differences in attenuation properties, size and shape along the scanned body region. As a consequence, the noise level of reconstructed images may vary along the *z*-direction. The automatic mA adaptation concept originated from the need to prevent patient over- or under-exposure, avoid photon starvation artifacts in high attenuation tissues and produce image series of approximately the same image quality (noise) irrespective of the position of the imaged slice along the *z*-axis (Lee 2008). Smart automatic mA modulation may prevent erroneous fixed mA selection by the operator that could result in non-diagnostic high-noise images when mA is set lower than the minimum required, or high radiation burden for the examined patient when mA is set much higher than the minimum required. The mA adaptation based on the total attenuation of transverse patient sections along the *z*-axis is referred to as *z*-axis modulation. *Z*-axis modulation results in different mA values for each location of the moving table along the *z*-axis during the scan. The mA adaptation based on the varying thickness of patient tissues that the CT beam has to penetrate as the CT beam rotates around the patient is referred to as angular mA modulation. Angular mA modulation results in differentiation of *z*-axis modulated values along the angular position of the CT tube around the patient.

All CT vendors have developed automatic mA modulation schemes that may have some minor differences between each other but the basic idea is the same. A preselected image quality level is preset and mA modulation aims to produce images of approximately the same image quality irrespective of the position along the *z*-axis.

In the GE Healthcare TCM scheme, the operator defines the desired image quality by selecting a parameter called 'noise index' (NI) and the minimum/maximum mA allowable values to avoid excessively low or high exposure of patient tissues, respectively. Setting NI high results in a low-dose acquisition at the cost of elevated noise in the resulting images. Selection of the appropriate NI for each examination is crucial since it determines the image quality level of a produced image series. Also, it is the primary determinant of the radiation dose received by the examined patient. Even if other exposure parameters that affect dose such as pitch and rotation time are changed, the tube current modulation scheme will adapt mA to achieve the prescribed image quality indicated by the NI setting. Based on the attenuation properties of the body region to be scanned and the selected NI and min-max mA, the mA-time curve is calculated prior to scanning and applied during scanning. To allow for both *z*-axis and angular mA adaptation, attenuation data of the region to be scanned are derived from a lateral and an anteroposterior radiographic scout view obtained prior to the scan.

In a similar fashion to GE Healthcare TCM, the Toshiba TCM system relies on setting the desired image quality by selecting a desired level of pixel value standard deviation (SD) and the minimum and maximum mA values in-between which the tube current is allowed to vary during acquisition. Selecting a higher SD for the target image series results in decreased dose at the cost of noisier images. Philips Healthcare and Siemens TCM systems are similar since a 'reference image quality' is employed to define the desired image quality of the target CT image series. In Philips

TCM the desired image quality is set through a previously acquired satisfactory patient CT examination. The attenuation data collected during scout view acquisition of the previous satisfactory study are compared to corresponding attenuation data collected for the patient to be examined through the scout views and mA is adapted to produce images of the same quality. Siemens TCM employs 'the reference effective mAs' (REmAs) to prescribe the desired image quality. The REmAs is actually the mAs value that produces satisfactory images for an average size adult (for adult CT studies) or an average size 5-year-old patient (for paediatric CT studies) adapted for the specific pitch value set for the patient to be examined. The size of the body region of the patient to be scanned is assessed through the scout view and compared to the reference patient to derive the mA modulation curve along the *z*-axis. Angular mA adaptation is achieved through a fast online feedback mechanism during rotation of the tube.

Setting the desired image quality for the target CT image series is prerequisite for all TCM systems. However, identifying the best choice for each patient examination is a challenging task. The minimum acceptable image quality level may vary even for the same CT examination type depending on the size of the patient, the clinical information expected from the examination and the experience of the radiologist. Moreover, the temporal modulation curve of mA during the scan is calculated using the attenuation data collected from the radiographic scout view image(s) always acquired prior to the scan for localization purposes. Since attenuation data for a patient to be examined are compared to corresponding data of a 'reference' patient through comparison of the corresponding scout view(s), mis-centering of the patient either in the vertical or the lateral axis may result in erroneous performance of the TCM system (Kaasalainen 2014). For all of the reasons above, wide knowledge of the basic underlying physics and familiarization with the specific characteristics of an TCM system is required by the operator if the benefits of TCM are to be fully exploited.

Some modern CT scanners allow organ-based tube current modulation (Wang *et al* 2012). This is referred to as the use of decreased mA value for a fraction of the 360° arc run by the X-ray tube at each rotation. The X-ray tube output may thus be reduced whenever the X-ray tube passes over radiosensitive organs/tissues that lay superficially in the anterior surface of the patient, such as eye-lenses, thyroid and breast. The achieved dose reduction in superficially located tissues has been reported to be 10%–30% (Duan *et al* 2011). However, in order to preserve adequate image quality in the resulting image series, the mA value should be increased for the remaining arc run by the rotating X-ray tube leading to increased dose to tissues that lay posteriorly (Gandhi *et al* 2015).

Cardiac CT acquisition mode

In cardiac CT studies the use of retrospective or prospective acquisition mode considerably affects the radiation dose to the examined patient (Hedgire *et al* 2017). Retrospective gating refers to continuous helical CT acquisition throughout the cardiac cycle. Collected attenuation data allow for reconstruction of images at any phase of the cardiac cycle at the expense of higher radiation burden for the patient. Prospective CT acquisition refers to a sequential CT acquisition mode synchronized with the electrocardiogram. The CT beam is turned on and off in synchronization with cardiac rhythm in order to collect attenuation data only during a fraction of the cardiac cycle. The beam is on only at the time between two successive cardiac pulses where the myocardium is almost unmoving (a time interval defined by the rotation time usually centered around the 70% of the entire R–R duration of the cardiac cycle). This is referred to as prospective ECG-gated mA modulation. Compared to retrospective, prospective ECG-gated acquisition may achieve >50% reduction of the patient radiation dose, with dose reduction being higher for fast CT scanners where low rotation time may be allowed for acquisition. However, prospective ECG-gated acquisition is prone to artifacts especially when the heart rate is >60 bpm or/and highly irregular. ECG-gated mA modulation may also be applied to retrospective cardiac CT acquisition. In this case, the beam is not turned off but a reduced value of mA is used for most of the cardiac cycle except for a preset time interval centered around the 70% of the cardiac cycle time where the myocardium is almost unmoving. Modern CT scanners allow ECG-gated mA modulation in either prospective or retrospective cardiac CT acquisition modes to efficiently minimize the radiation burden for the examined patient.

Automatic tube voltage selection

Following many research publications on the benefits of using lower kV values for CT examinations traditionally performed at 120 kV, some manufacturers have developed automatic tube voltage selection (ATVS) schemes. The optimum kV value is determined on the basis of patient size and the specific diagnostic task. Information regarding the size of the patient is obtained from the scout views. ATVS system may be used in conjunction with TCM (Lee 2012). Having defined the optimum kV selection, the most appropriate TCM parameters are proposed to achieve images of desired image quality. ATVS systems provide a 'wise' selection of tube voltage that might not be always reached by the operator whatever is his/her experience. ATVS systems have been reported to be particularly advantageous regarding minimization of CT exposure in paediatric and small-size adult patients.

Over-scanning and adaptive section collimation

In helical scanning the imaged volume boundaries do not coincide with the boundaries of the primarily exposed volume as in sequential CT scanning. This is due to the need for acquiring attenuation data also for patient body regions that lie outside the prescribed volume to be imaged. Namely, interpolation algorithms employed in helical CT to reconstruct CT images require attenuation data from several neighboring patient slices at either side of the prescribed body region to be imaged. This is called over-scanning (or over-ranging) and has been reported to considerably increase the dose to the patient when helical instead of sequential scanning is performed (Tzedakis *et al* 2005). The over-scanning extent at either side of the prescribed volume to be imaged increases with beam collimation, pitch and reconstructed slice thickness. If however, all these parameters are held constant the over-scanned region has the same length. The ratio of over-scanning extent over the prescribed image volume is decreased as the image volume length increases. Consequently, the %difference in patient radiation dose burden between helical and sequential CT acquisition when all other exposure parameters are held constant depends on the prescribed image volume length along the *z*-axis. The shorter the volume to be imaged along the *z*-axis, the higher the %dose increase due to over-scanning when helical instead of sequential scanning is employed. Dose differences between helical and sequential acquisition of common CT examinations has been reported to be 10%–40% for adult patients and may be considerably higher for children (Tzedakis *et al* 2006) where the imaged volume length along the *z*-axis is shorter due to the minor body size.

Most modern CT scanners are equipped with an adaptive section collimation system that aims to reduce the extra dose received by the patient due to over-scanning (Deak *et al* 2009). This system reduces the extent of primarily exposed patient tissues that lie outside the prescribed image volume, thus achieving a substantial dose saving for the examined patient.

Iterative reconstruction algorithms

The traditional method for reconstruction of CT images from attenuation data collected during a CT scan is the filtered back-projection (FBP) method. FBP does not require high computing power and may generate a CT image series in a few seconds following acquisition of CT attenuation data. Being the only practical reconstruction method for several decades after the advent of CT, FBP has dominated CT imaging until recently. FBP, however, relies on simplified assumptions and approximations and therefore it is prone to beam hardening and scatter effects that may result in streak artifacts and high noise, especially in low-dose acquisitions. Besides, algebraic methods for CT image reconstruction had been described and proposed in the early years of CT. Algebraic reconstruction methods were reported to allow considerable suppression of image artifacts and noise for the same level of patient exposure. However, they required immense computational power that rendered them impractical for use in clinical environments for several decades. Computational power rose exponentially during the 1990s and 2000s allowing algebraic methods to be timidly included as an option in many commercial CT systems. Over the last decade, CT vendors have invested a high level of resources to develop advanced iterative reconstruction (IR) methods. Nowadays, modern CT suites provide very efficient IR tools that have been reported to achieve much higher image quality at reduced patient dose levels in comparison to FBP (Geyer 2015). Substantial dose savings well exceeding 50% have been reported when an IR instead of FBP reconstruction algorithm is used, especially in average- and large-size patients. Currently, the use of the IR algorithm is considered to be the most efficient dose sparing tool available in modern CT scanners. However, modern IR methods are highly complex and therefore high familiarization with the underlying physics and mathematics is required from users to achieve full exploitation of their benefits.

3.2.3 Patient-related factors affecting radiation dose to conceptus or children exposed during a CT examination

Patient size

The radiation dose to exposed organs/tissues of a patient subjected to CT exposure depends on patient body size. In general, the organs/tissues of a slim patient receive much higher amounts of dose compared to a large patient if both patients are subjected to identical CT exposures, as illustrated in figure 3.5. The radiosensitive organs of a large patient are exposed to less radiation intensity due to the increased attenuation of overlying tissues. Therefore, exposure parameters for slim adult patients and children should be properly adjusted to prevent excessive amounts of radiation dose absorbed by patient tissues during CT exposure.

The dose-length product (DLP), provided by the CT system as an indicator of radiation dose to the examined patient, is commonly used to derive an estimate of the effective dose (ED) to the patient through CT examination-specific DLP-to-ED conversion factors. However, patient size should be encountered by applying size-specific conversion factors otherwise ED estimates are rough approximations (Perisinakis *et al* 2008, Deak *et al* 2010).

Imaged volume length along the z-axis

The boundaries of a patient's body region to be imaged during a CT examination are commonly set by the operator on the scout view(s) always acquired prior to a CT

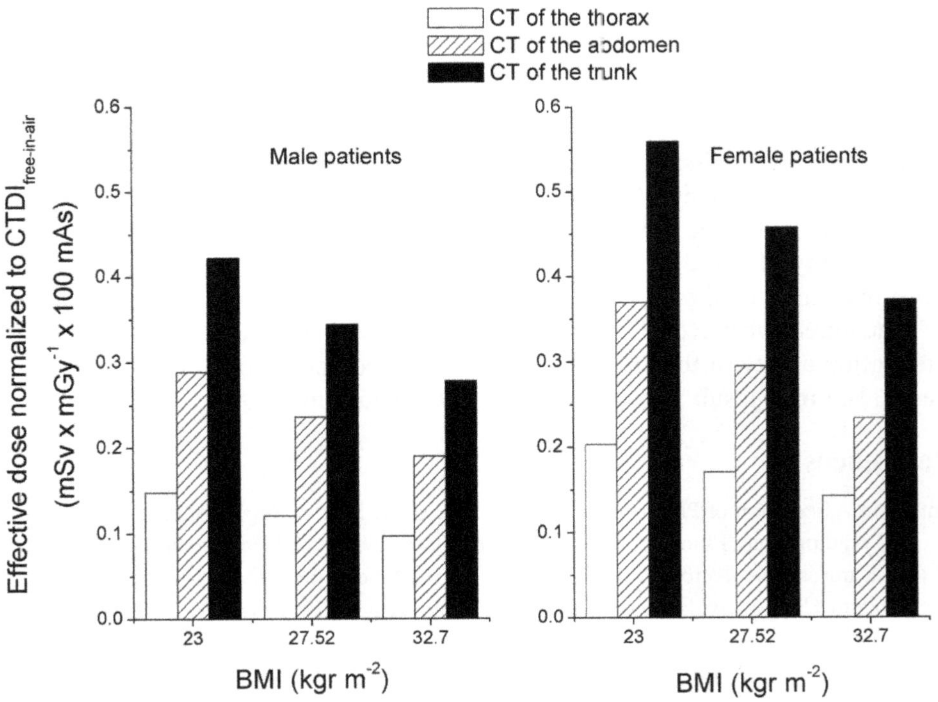

Figure 3.5. The effect of patient size on dose from a CT exposure with the same exposure parameters. From Perisinakis *et al* (2008).

scan. Specific anatomical landmarks are used to define the extent of image volume in standardized CT examinations. Apparently, the longer the primarily exposed body region along the *z*-axis, the higher the radiation dose burden for the examined patient. Effective dose from a CT examination is approximately proportional to prescribed scan length along the *z*-axis. In general, the boundaries of the image volume should be set with caution to include only the body region under investigation.

The larger/taller the patient, the longer the body part to be imaged in a CT examination. Namely, for a chest CT examination the imaged volume length along the *z*-axis for an adult patient may be set above 40 cm while in a neonate may be as small as 15 cm. Image volume extent along the *z*-axis is the main determinant of the scanning length, i.e. the length of the patient's body region to be primarily exposed during acquisition. Apart from the prescribed image volume length along the *z*-axis, scanning length is affected by the over-scanning effect discussed in section 3.2.2.

Patient tissues that lie outside the primarily exposed body region may absorb dose due to scatter. The amount of dose absorbed by such tissues declines with the distance from boundaries of the primarily exposed body region. In the case of CT examinations on pregnant patients where the embryo tissues lie outside or partly inside the primarily exposed body region, setting the imaged volume to the minimum required may reduce the radiation dose to both the expectant mother and the embryo.

In-plane shielding of superficial radiosensitive tissues

In order to reduce the dose from CT exposure absorbed by radiosensitive tissues that lay superficially in the human body, the use of radioprotective garments made by high-z material has been proposed (Wang *et al* 2012). These garments are positioned in-plane to reduce the dose to tissues primarily exposed during acquisition. Eye-lens, thyroid and breast tissue radioprotective garments have been reported to reduce dose to these organs by 20%–40%. However, radioprotective garments should be used with caution since they may induce beam hardening artifacts degrading CT image quality (Raissaki *et al* 2010). Also, it has to be stressed that the radio-protective garment should be positioned on the patient after the scout view(s) has been acquired whenever TCM is employed. If not, incorrect data regarding the attenuation ability of the imaged tissues is obtained and automatic mA modulation mechanism may result in erroneously high tube current.

References

American Association of Physicists in Medicine, (AAPM) 2012 A Report of AAPM Task Group 125 Radiography/Fluoroscopy Subcommittee, Science Council Functionality and Operation of Fluoroscopic Automatic Brightness Control/Automatic Dose Rate Control Logic in Modern Cardiovascular and Interventional Angiography Systems

Bacher K, Bogaert E and Lapere R *et al* 2005 Patient-specific dose and radiation risk estimation in pediatric cardiac catheterization *Circulation* **111** 83–9

Beck T J and Rosenstein M 1979 *Quantification of current practice in pediatric roentgenography for organ dose calculations* HEW Publication (FDA) 79–8078. US Department of Health

Education and Welfare, Public Health Service, Food and Drug Administration (Rockville, Maryland: Bureau of Radiological Health), pp 34

Berrington de Gonzalez A, Mahesh M and Kim K P *et al* 2009 Projected cancer risks from computed tomographic scans performed in the United States in 2007 *Arch. Intern. Med.* **169** 2071–7

Boone J M, Pfeiffer D E, Strauss K J, Rossi R P, Lin P, Shepard J and Conway B 1993 A survey of fluoroscopic exposure rates: AAPM Task Group No 11 Report *Med. Phys.* **20** 789–94

Brosi P, Stuessi A, Verdun F R, Vock P and Wolf R 2011 Copper filtration in pediatric digital X-ray imaging: its impact on image quality and dose *Radiol. Phys. Technol.* **4** 148–55

Carlton R R and Adler A M 1996 *Principles of Radiographic Imaging* 2nd edn (Albany: Delmar), p 266

Council of the European Communities 1996 *European guidelines on quality criteria for diagnostic radiographic images in paediatrics* Report EUR 16261 (EN, Luxemburg: European Commission)

Cowen A R, Davies A G and Sivananthan M 2008 The design and imaging characteristics of dynamic, solid state, flat-panel x-ray image detectors for digital fluoroscopy and fluorography *Clin. Radiol.* **63** 1073–85

Damilakis J, Perisinakis K, Prassopoulos P, Dimovasili E, Varveris H and Gourtsoyiannis N 2003 Conceptus radiation dose and risk from chest screen-film radiography *Eur Radiol.* **13** 406–12

Damilakis J, Theocharopoulos N, Perisinakis K, Manios E, Dimitriou P, Vardas P and Gourtsoyiannis N 2001 Conceptus radiation dose and risk from cardiac catheter ablation procedures *Circulation* **104** 893–7

Damilakis J, Theocharopoulos N, Perisinakis K, Papadokostakis G, Hadjipavlou A and Gourtsoyiannis N 2003 Conceptus radiation dose assessment from fluoroscopically assisted surgical treatment of hip fractures *Med. Phys.* **30** 2594–601

Deak P D, Langner O and Lell M *et al* 2009 Effects of adaptive section collimation on patient radiation dose in multisection spiral CT *Radiology* **252** 140–7

Deak P D, Smal Y and Kalender W A 2010 Multisection CT protocols: sex- and age-specific conversion factors used to determine effective dose from dose-length product *Radiology* **257** 158–66

Dragusin O, Gewillig M and Desmet W *et al* 2008 Radiation dose survey in a paediatric cardiac catheterization laboratory equipped with flat-panel detectors *Radiat. Prot. Dosim.* **129** 91–5

Duan X, Wang J, Christner J A, Leng S, Grant K L and McCollough C H 2011 Dose reduction to anterior surfaces with organ-based tube-current modulation: evaluation of performance in a phantom study *AJR Am. J. Roentgenol.* **197** 689–95

Food and Drug Administration (FDA) 2017 *Pediatric Information for X-ray Imaging Device Premarket Notifications.* U.S. Department of Health and Human Services, Center for Devices and Radiological Health

Freitas M B and Yoshimura E M 2004 Dose measurements in chest diagnostic X rays: adult and paediatric patients *Radiat. Prot. Dosim.* **111** 73–6

Gagne R and Quinn P 1995 X-ray spectral considerations in fluoroscopy *1995 Syllabus: Categorical Course in Physics-Physical and Technical Aspects of Angiography and Interventional Radiology* ed S Balter and T Shope (Oak Brook, IL: Radiological Society of North America), pp 49–58

Gandhi D, Crotty D J, Stevens G M and Schmidt T G 2015 Technical note: phantom study to evaluate the dose and image quality effects of a computed tomography organ-based tube current modulation technique *Med. Phys.* **42** 6572–8

Geyer L L, Schoepf U J and Meinel F G *et al* 2015 State of the art: iterative CT reconstruction techniques *Radiology* **276** 339–57

Goldberg-Stein S, Liu B and Hahn P F *et al* 2011 Body CT during pregnancy: utilization trends, examination indications, and fetal radiation doses *Am. J. Roentgenol.* **196** 146–51

Harrison R M 1981 Central axis depth dose data for diagnostic radiology *Phys. Med. Biol.* **26** 657–70

Hedgire S, Ghoshhajra B and Kalra M 2017 Dose optimization in cardiac CT *Phys. Med.* **41** 97–103

ICRP 2001 ICRP publication 85: avoidance of radiation injuries from medical interventional procedures, *Ann. ICR.* **30**

ICRP 2007 The 2007 Recommendations of the International Commission on Radiological Protection. ICRP Publication 103 *Ann. ICRP* **37** 1–332

International Commission on Radiological Protection 1975 *Report of the Task Group on Reference Man* (Oxford: Pergamon) ICRP publication 23

Kaasalainen T, Palmu K and Reijonen V *et al* 2014 Effect of patient centering on patient dose and image noise in chest CT *Am. J. Roentgenol.* **203** 123–30

Kocher K E, Meurer W J and Fazel R *et al* 2011 National trends in use of computed tomography in the emergency department *Ann. Emerg. Med.* **58** 452–62

Kubo T 2019 Vendor free basics of radiation dose reduction techniques for CT *Eur. J. Radiol.* **110** 14–21

Lazarus E, Debenedectis C, North D, Spencer P K and Mayo-Smith W W 2009 Utilization of imaging in pregnant patients: 10-year review of 5270 examinations in 3285 patients – 1997–2006 *Radiology* **251** 517–24

Lee C H, Goo J M and Ye H J *et al* 2008 Radiation dose modulation techniques in the multidetector CT era: from basics to practice *Radiographics* **28** 1451–9

Lee K H, Lee J M and Moon S K *et al* 2012 Attenuation-based automatic tube voltage selection and tube current modulation for dose reduction at contrast-enhanced liver CT *Radiology* **265** 437–47

Maldjian P D and Goldman A R 2013 Reducing radiation dose in body CT: a primer on dose metrics and key CT technical parameters *Am. J. Roentgenol.* **200** 741–7

Martin C 2007 The importance of radiation quality for optimisation in radiology *Biomed. Imaging Interv. J.* **3** e38

McCollough C H, Schueler B A, Atwell T D, Braun N N, Regner D M and Brown D L *et al* 2007 Radiation exposure and pregnancy: when should we be concerned? *Radiographics* **27** 909–17 Discussion 917–8

Omnasch D, Schemm A and Kramer H 2004 Optimization of radiographic parameters for pediatric cardiac angiography *Br. J. Radiol.* **77** 479–87

Perisinakis K, Damilakis J, Vagios E and Gourtsoyiannis N 1999 Embryo depth during the first trimester. Data required for embryo dosimetry *Invest. Radiol.* **34** 449–54

Perisinakis K, Papadakis A and Damilakis J 2009 The effect of x-ray beam quality and geometry on radiation utilization efficiency in multi-detector CT imaging *Med. Phys.* **36** 1258–66

Perisinakis K, Seimenis I and Tzedakis A *et al* 2013 The effect of head size/shape, miscentering and bow-tie filter on peak patient tissue doses from modern brain perfusion 256-slice CT: How can we minimize the risk for deterministic effects? *Med. Phys.* **40** 011911

Perisinakis K, Tzedakis A and Damilakis J 2008 On the use of Monte Carlo-derived dosimetric data in the estimation of patient dose from CT examinations *Med. Phys.* **35** 2018–28

Pola A, Corbella D and Righini A *et al* 2018 Computed tomography use in a large Italian region: trend analysis 2004-2014 of emergency and outpatient CT examinations in children and adults *Eur. Radiol.* **28** 2308–18

Ragozzino M W, Gray J E, Burke T M and Van Lysel M S 1981 Estimation and minimization of fetal absorbed dose: data from common radiographic examinations *Am. J. Roentgenol.* **137** 667–71

Raissaki M, Perisinakis K, Damilakis J and Gourtsoyiannis N 2010 Eye-lens bismuth shielding in paediatric head CT: artefact evaluation and reduction *Pediatr. Radiol.* **40** 1748–54

Samara E T, Stratakis J, Enele Melono J M, Mouzas I A, Perisinakis K and Damilakis J 2009 Therapeutic ERCP and pregnancy: is the radiation risk for the conceptus trivial? *Gastrointest. Endosc.* **69** 824–31

Shrimpton P C, Jones D G and Wall B F 1988 The influence of tube filtration and potential on patient dose during X-ray examinations *Phys. Med. Biol.* **33** 1205–12

Stovall M, Blackwell C R, Cundiff J, Novack D H, Palta J R, Wagner L K, Webster E W and Shalek R J 1995 Fetal dose from radiotherapy with photon beams: report of AAPM Radiation Therapy Committee Task Group No. 36 *Med. Phys.* **22** 63–82 Review

Theocharopoulos N, Damilakis J, Perisinakis K, Papadokostakis G, Hadjipavlou A and Gourtsoyiannis N 2006 Fluoroscopically assisted surgical treatments of spinal disorders: conceptus radiation doses and risks *Spine* **31** 239–44

Tirada N, Dreizin D, Khati N J, Akin E A and Zeman R K 2015 Imaging pregnant and lactating patients *Radiographics* **35** 1751–65

Tzedakis A, Damilakis J, Perisinakis K, Stratakis J and Gourtsoyiannis N 2005 The effect of z-overscanning on patient effective dose from multi-detector spiral computed tomography examinations *Med. Phys.* **32** 1621–9

Tzedakis A, Perisinakis K and Raissaki M *et al* 2006 The effect of z-overscanning on radiation burden of pediatric patients undergoing head CT with multi-detector scanners. A Monte Carlo study *Med. Phys.* **33** 2472–8

Wang J, Duan X, Christner J A, Leng S, Grant K L and McCollough C H 2012 Bismuth shielding, organ-based tube current modulation, and global reduction of tube current for dose reduction to the eye at head CT *Radiology* **262** 191–8

Yadava G K, Kuhls-Gilerist A, Rudin S, Patel V, Hoffmann K and Bednarek D 2008 A practical exposure-equivalent metric for instrumentation noise in x-ray imaging systems *Phys. Med. Biol.* **53** 5107–21

Yakoumakis E, Kostopoulou H, Makri T, Dimitriadis A, Georgiou E and Tsalafoutas I 2013 Estimation of radiation dose and risk to children undergoing cardiac catheterization for the treatment of a congenital heart disease using Monte Carlo simulations *Pediatr. Radiol.* **43** 339–46

Yakoumakis E N *et al* 2007 Radiation doses in common X-ray examinations carried out in two dedicated paediatric hospitals *Radiat. Prot. Dosimetry* **124** 348–52

Zamenhof R G 1982 The optimization of signal detectability in digital fluoroscopy *Med. Phys.* **9** 688–94

IOP Publishing

Radiation Dose Management of Pregnant Patients, Pregnant Staff and Paediatric Patients
Diagnostic and interventional radiology
John Damilakis

Chapter 4

Amount of dose absorbed by the conceptus and paediatric patients from diagnostic and interventional radiology

J Damilakis and V Tsapaki

This chapter is divided into two main parts. First, information is provided about the amount of radiation dose absorbed by the conceptus and radiation-induced risks from diagnostic and interventional X-ray examinations. The second part deals with radiation doses and radiogenic risks associated with paediatric diagnostic and fluoroscopically guided procedures. This information is important for the justification as well as for the optimization of the exposure.

4.1 Conceptus dose and radiation-induced risk associated with diagnostic and interventional X-ray examinations

Radiation exposure of pregnant patients undergoing diagnostic or interventional X-ray examinations is of scientific and social concern. The risk–benefit consideration for radiological evaluation of the patient is sometimes complicated as radiogenic risks to both mother and unborn child have to be taken into consideration. Moreover, risks to the mother should also be considered if the examination is postponed or avoided. Knowledge on the amount of radiation dose absorbed by the conceptus from diagnostic and interventional X-ray examinations and associated radiogenic risk is important because this information is very useful for the justification as well as for the optimization of the exposure. Of note, however, is that there is a large variation in conceptus dose from the same type of examination among published studies. This variation reflects the use of different equipment and techniques in the institutions where studies were performed. The following paragraphs provide published typical conceptus doses from diagnostic and interventional procedures. The actual value of conceptus dose from a specific examination may vary considerably depending upon many factors such as the exposure parameters applied for the performance of the examination, reconstruction algorithms, use of dose reduction tools and the body size of the patient. For this reason, accurate estimation of the conceptus dose is needed for the evaluation of a particular patient case using a method developed specifically for this purpose. Information about methods available for the estimation of conceptus radiation dose can be found in chapter 5.

4.1.1 Conceptus dose and radiogenic risk associated with radiography

Radiographic or fluoroscopic evaluation of pregnant patients may be required at any stage of gestation. Trauma and acute abdominal or thoracic pain are the leading conditions where X-ray imaging is needed for pregnant patients. Chest radiographs are the most frequently performed X-ray examinations. Conceptus dose from chest radiography examinations carried out in pregnant women in all trimesters of gestation has been reported in the literature. Damilakis *et al* (2003a) have found that the anteroposterior (AP) chest radiographs performed on a pregnant patient with a chest size of 22.3 cm during the first trimester result in conceptus doses ranging between 0.0021 and 0.0028 mGy (Damilakis *et al* 2003a). Figure 4.1 (left) shows conceptus doses from an AP chest radiograph carried out on a patient during the first postconception weeks as a function of maternal chest thickness and three conceptus depths (5.5 cm, 8.5 cm and 11.5 cm). Corresponding doses to the unborn child at the second and third trimesters are higher than that of the first trimester but normally do not exceed 0.15 mGy (figure 4.1 (right)) even for patients with large chest thickness. Conceptus dose from a pulmonary angiography (PA) chest radiograph are lower than those from an AP projection. Conceptus radiogenic risk can be calculated using risk factors recommended by radiation protection authorities. Of note, however, is that considerable uncertainty exists about risks calculated using these factors since they are derived from epidemiologic studies on the effects of individuals exposed to high levels of ionizing radiation, mainly from atomic bomb survivors and patients exposed during radiotherapy. In the study by Damilakis *et al* (2003a), the radiogenic risk of fatal cancer was calculated by multiplying the

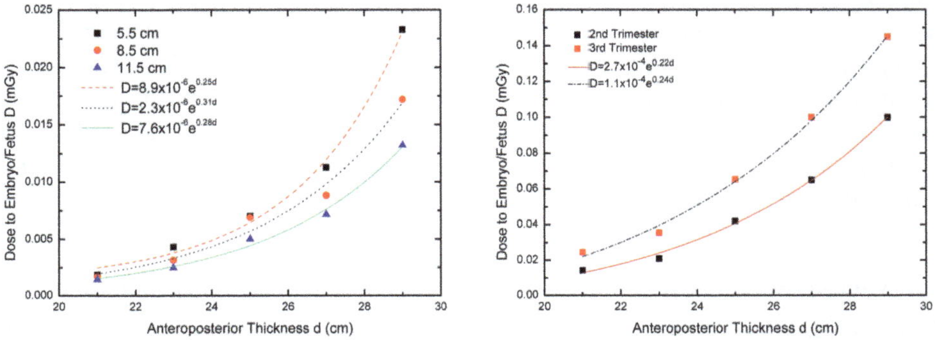

Figure 4.1. Conceptus dose from an AP chest radiograph during the first trimester (left) and during the second and third trimesters (right).

conceptus dose with a risk coefficient of 10^{-1} Gy^{-1} recommended by the National Council on Radiation Protection and Measurements (NCRP 1994). The risk factor for induction of hereditary effects was considered to be equal to 10^{-2} Gy^{-1}. The maximum conceptus risk of fatal childhood cancer per chest projection was found to be 10.7×10^{-6} for an AP chest radiograph performed on a patient with chest thickness of 27.2 cm during the last months of gestation. The natural incidence rate of fatal childhood cancer is 769×10^{-6}. Therefore, the conceptus risk per AP chest projection is very low in comparison with the natural cancer risk. The maximum risk of hereditary effects in future generations was found 1.1×10^{-6} for an AP chest radiograph performed on a mother with chest thickness of 27.2 cm during the third trimester. The current incidence of birth defects is about 6%. Therefore, the increased genetic risk for a conceptus associated with the highest dose from a chest radiograph is considerably smaller compared with the natural frequency of genetic disease. The above study also showed that the use of a lead apron does not decrease the conceptus dose significantly. Proper selection of field size is of great importance in chest radiography in late pregnancy, where the distance between the conceptus and the edge of the X-ray field is limited. Among the most important parameters that affect conceptus doses in chest radiography are the tube potential and the filtration of the X-ray tube. Increase in kVp and adequate filtration would yield a reduction in conceptus dose without loss of image quality (IQ). Conceptus doses can also be minimized by selecting PA instead of AP projection.

Osei and Faulkner estimated the dose to the unborn child of 50 pregnant patients over a period of 10 years (Osei and Faulkner 1999). They provide detailed information about the range and the mean conceptus dose from various radiographic studies. As expected, radiographic examinations performed in the abdominopelvic region (abdomen, lumbar spine, pelvis) are associated with the highest conceptus dose (mean conceptus dose up to 7.5 mGy for AP lumbar spine) whereas extra-abdominal studies deliver considerably lower doses to the unborn child (mean conceptus doses lower than 0.01 mGy for all chest and thoracic spine projections).

Dual X-ray absorptiometry (DXA) measurements are occasionally performed during pregnancy. These measurements may identify pregnancy-associated osteoporosis and exclude diseases with clinical features similar to those of osteoporosis. A study showed that the embryo/fetus doses during all trimesters of gestation are on the order of a few μGys (Damilakis *et al* 2002). The maximum conceptus dose during the first trimester associated with DXA examinations was 3.4 μGy due to a proximal femur scan. The conceptus doses during the second and third trimesters were 2.7 and 4.9 μGy, respectively, for the scans of the lumbar spine and 1.4 and 1.0 μGy for the proximal femur exams. These doses are negligible and DXA examinations to pregnant patients should not be postponed due to radiation dose.

4.1.2 Conceptus dose and radiogenic risk associated with computed tomography (CT)

The radiation dose to the embryo or fetus from CT examinations performed on the mother depends on many parameters including exposure parameters (kV, mA, sec), pitch, beam collimation, image slice thickness, scan field-of-view, scanning length, scout views, patient centering, type of scan (axial/helical), the use of dose reduction tools (automatic exposure control, automatic kV selection, organ dose modulation and tools used to reduce z-overscanning), reconstruction algorithm, number of phases, scanning length, the use of contrast material, the body size of the mother, the trimester of gestation and the depth of the conceptus. As a rule of thumb, direct conceptus exposure during abdominopelvic CT studies is associated with relatively high doses. In contrast, a conceptus dose from indirect exposure such as exposure from head or chest CT has not been shown to exceed 1 mGy.

Physical phantoms and dose meters as well as Monte Carlo simulation codes and mathematical phantoms have been used to estimate embryo/fetal doses from various protocols. Mathematical phantoms of the pregnant patient and her baby have been available for many years. Stylized phantoms of pregnant patients were developed by Stabin *et al* (1995) representing pregnancy at the end of each trimester of gestation. Xu and colleagues have also developed mathematical phantoms at the end of the three gestational periods (Xu *et al* 2007). These phantoms were used to estimate conceptus doses from abdominal CT (Gu *et al* 2013). Three tube current modulation (TCM) schemes clinically used for pregnant patients were utilized to take into consideration the effect of TCM in radiation exposure. When TCM was applied, conceptus doses ranged from 7.8 to 20.9 mGy compared with 9.4–28.0 mGy when TCM was deactivated.

Only a few research groups have constructed physical anthropomorphic phantoms to estimate conceptus doses from CT protocols. Home-made gelatin boluses to simulate pregnancy have been used in a study by Kelaranta *et al* (2017). The authors used these phantoms and metal-oxide semiconductor field-effect transistor (MOSFET) dosimeters to estimate conceptus doses from pulmonary angiography, abdominopelvic and trauma CT protocols. MOSFET dose values were compared with conceptus dose estimates derived from Monte Carlo simulation experiments. Conceptus doses in different stages of pregnancy were found to be up to 1.04 mGy, up to 5.8 mGy, and up to 12.6 mGy for pulmonary angiography, abdominopelvic and trauma CT scans, respectively. Damilakis *et al* (2000) modified the Alderson Rando anthropomorphic phantom to simulate late pregnancy. Seven and ten rings of Lucite were added to Rando slices to simulate pregnancy in the second and third trimesters of gestation, respectively. Each ring contained holes that allowed placement of TLDs. These phantoms have been used to estimate conceptus abdominal doses from CT during the second and third trimesters of gestation. Radiation doses of up to 43.6 mGy were measured in the measuring locations of these phantoms.

Pulmonary embolism is the leading non-obstetric cause of maternal death. Prevalence of thromboembolism in pregnant and postpartum women is two to four times greater than that for nonpregnant patients of the same age group. CT pulmonary angiography (CTPA) and lung perfusion scintigraphy are the most frequently used imaging methods for the evaluation of pulmonary embolism. According to international recommendations, chest radiography is recommended as the first X-ray imaging procedure. If the results are normal, lung perfusion scintigraphy should be performed instead of CTPA because CT delivers a significantly higher dose to the patient than lung perfusion scintigraphy. If the results of chest radiography are abnormal, CT pulmonary angiography is recommended as the next step of investigation, which is the reference standard for the diagnosis of pulmonary embolism.

A suspected pulmonary embolism is among the clinical conditions where direct exposure of the conceptus from diagnostic X-rays is not needed. As a consequence, conceptus doses are very low because the distance from the directly irradiated region to the location of the conceptus is at least 20 cm. A study (Perisinakis *et al* 2014) showed that a typical conceptus dose from low-dose protocols is lower than 0.3 mGy. Figure 4.2 shows the dose to the unborn child from low-dose CTPA performed on pregnant patients of varying body size and gestational stage as well as corresponding data for lung perfusion scintigraphy. The risk of childhood fatal cancer was estimated by multiplying the corresponding dose by a risk factor of 6% per Gy recommended by ICRP (2000). Risks were lower than 2×10^{-5} for all trimesters of gestation. An important conclusion of this study is that lung perfusing scintigraphy is more dose-efficient in comparison with CTPA even when modern CT technology is used.

Appendicitis is the most common non-obstetrical indication for surgical exploration of the maternal abdomen. When imaging is required, ultrasound is the modality of choice. If the ultrasound is negative or equivocal it may be more appropriate to proceed to exploratory laparotomy than to perform abdominal CT. However, if surgical exploration is contraindicated, CT may be considered. CT is highly sensitive (up to 98%) and specific (up to 97%) for the diagnosis of acute appendicitis and

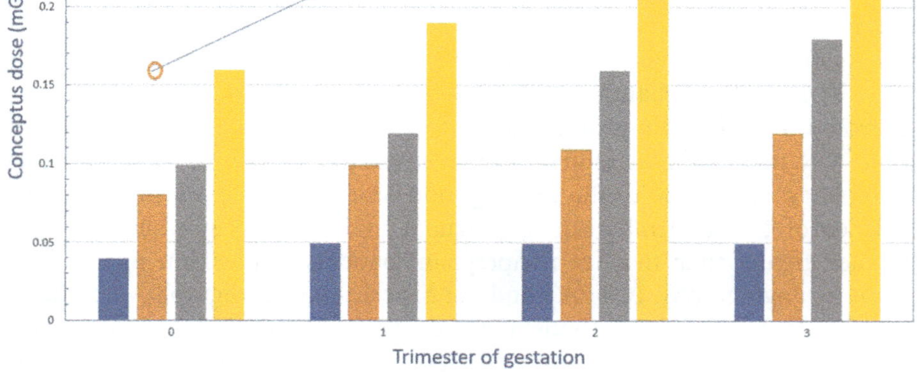

Figure 4.2. Conceptus dose from low-dose CTPA performed using a 256-slice CT scanner is lower than the corresponding dose from low-dose lung perfusion scintigraphy, with the exception of pregnant patients with a body mass index (BMI) greater than 30 kg m^{-2} at the end of the third trimester of gestation.

allows for differential diagnosis of abdominal pain. A standard CT protocol for appendicitis would result in a conceptus dose up to 30 mGy. Low-dose protocols can reduce the dose to the embryo or fetus considerably.

Urinary calculus disease is the most common painful condition requiring admission of the pregnant patient to hospital. Ultrasound can diagnose the obstruction and may disclose the obstructing stone. When the ultrasound is inconclusive, MR urography and low-dose CT have to be considered. Urinary calculus evaluation can be performed by using lower radiation doses than those used in evaluation of appendicitis because detection of high-contrast structures is affected less by high image noise than low-contrast structures. White *et al* (2007) have published their experience using low-dose CT in the evaluation of pregnant patients with refractory flank pain. This study found that CT was more sensitive in locating urinary calculi than renal US. Using modern CT technology, conceptus dose from a low-dose CT protocol for the urinary calculus evaluation can be lower than 10 mGy.

Injuries occur in the course of 6%–7% of all pregnancies with a highest frequency during the third trimester of gestation. Automobile accidents and physical abuse are the most common causes. Although less than 0.5% of the traumatized pregnant women need hospitalization, approximately one fourth of them with major injuries fail to survive. Trauma is the leading non-obstetric cause of maternal mortality accounting for 22% of deaths with head and abdominal injuries more frequently responsible. CT is considered a very beneficial high-dose emergency radiology technique. When the conceptus is exposed directly, its dose is in the range of 10–40 mGy and can be higher in exceptional circumstances. If the conceptus is not in the view of the beam, radiation dose is unlikely to exceed 1 mGy. In polytrauma cases, several anatomical regions may be scanned and the contribution from each scan series should be determined in order to estimate the total dose to the unborn child.

4.1.3 Conceptus dose and radiogenic risk associated with fluoroscopy and fluoroscopically guided procedures

Only a few studies have investigated the dose to the unborn child from fluoroscopic studies performed on the mother. Damilakis *et al* (1996) estimated embryo doses from barium enema examinations in case of accidental exposure. Doses ranged from 19 to 81 mGy and exceeded 50 mGy only when fluoroscopy time was longer than 7 min.

The number of fluoroscopically guided procedures has increased considerably over the last few years. Occasionally, pregnant patients are exposed to ionizing radiation from these procedures, intentionally or accidentally. Radiation exposure from interventional procedures may be relatively high and, for this reason, pregnant patients undergoing these procedures require specific consideration.

Some examinations such as transureteral stent, uterine fibroid embolization and CT-guided interventions are associated with high conceptus doses especially if the unborn child is exposed primarily to the X-ray beam. Metzger and Van Riper have estimated the conceptus dose from various fluoroscopically guided procedures (Metzger and Van Riper 1999). They found that the conceptus dose from placement of a transureteral stent was 44 mGy. When conceptus doses are suspected of exceeding 10 mGy, installation-specific measurements and calculations are needed for an accurate estimation of conceptus dose (ICRP 2000).

Endoscopic retrograde cholangiopancreatography (ERCP) has been used for the diagnosis and treatment of pancreatic diseases. Samara *et al* (2009) estimated conceptus doses and risks from therapeutic ERCP procedures. The radiogenic risk of fatal childhood cancer was calculated by multiplying the conceptus dose with a risk factor of 3×10^{-2} per Gy. The risk coefficient for induction of hereditary effects was considered to be equal to 2.4×10^{-2} per Gy. Conceptus doses ranged from 3.4 to 55.9 mGy. Authors found that the dose to the unborn child may exceed 10 mGy when the total dose-area product (DAP) of the procedure is higher than 130 mGy cm^2. Risk for fatal childhood cancer ranged from 10^{-4} to 1.7×10^{-3}. The risk for hereditary effects in future generations ranged from 8.2×10^{-5} to 1.3×10^{-3}.

Many young patients receive fluoroscopically guided surgical treatments of the spine such as pedicle screw fixation, vertebroplasty and kyphoplasty. Accidental exposure of the conceptus to radiation is always possible. Conceptus doses from typical fluoroscopically guided surgical treatments of the spine at five vertebral levels are provided by Theocharopoulos *et al* (2006). Conceptus doses from fluoroscopically assisted surgical treatment of spinal disorders are lower than 4 mGy during all gestational ages provided that the conceptus lies outside the primarily irradiated region. When the conceptus is primarily irradiated, the mean dose can exceed 100 mGy in very rare cases.

In comparison with several other X-ray studies, relatively high doses are delivered to patients undergoing radiofrequency cardiac catheter ablation procedures. High doses result from the prolonged fluoroscopy time often required for these studies; sometimes the total fluoroscopy time is longer than 1 h. Nevertheless, for a typical examination performed on a pregnant patient, radiation dose to the conceptus is less than 1 mGy in all periods of gestation (Damilakis *et al* 2001). Conceptus dose is so low because during typical ablation procedures, the conceptus is in the direct path (primary X-ray beam) of the X-ray beam for only 20–30 s that is at the beginning of the procedure, when the electrode catheters are inserted and advanced from the groin to the heart under fluoroscopic guidance in the posteroanterior projection.

Screening time as long as 14 min has been reported in the literature for surgical treatment of hip fractures. The projections involved in these procedures are (a) PA and (b) lateral cross-table 45°. Although the path of the X-ray beam is in the proximity of the unborn child the dose to the conceptus is less than 1 mGy during all trimesters of gestation because the embryo or fetus is not exposed primarily to radiation (Damilakis *et al* 2003b).

4.2 Paediatric dose and radiation-induced risk associated with diagnostic and interventional X-ray examinations

There are specific documents that deserve special attention related to patient dose in paediatric patient dose:

1. *IAEA Human Health Series No. 24 (2014). Dosimetry in Diagnostic Radiology for Paediatric Patients.* This publication draws on an IAEA

coordinated research project and provides recommendations specific to the measurement and interpretation of radiation dose to children received as a result of undergoing diagnostic radiological examinations. It complements the work of the Technical Reports Series No. 457 on Dosimetry in Diagnostic Radiology: An International Code of Practice and extends this work in methodologies for dosimetry in clinical environments to that required for non-adult patients. It includes dosimetry methodologies for general radiography, fluoroscopy and CT for both phantom and patient measurements. Details are given on dose audit strategies that take into account the size of children and on how the results of such audits can be used to indicate or be related to diagnostic reference levels. The effects of radiation on non-adults are also reviewed, as are the factors involved in the management of paediatric dosage in the clinical setting.

2. *IAEA Safety Reports Series No. 71 (IAEI 2017). Radiation Protection in Paediatric Radiology.* This publication provides guidance to radiologists, other clinicians and radiographers/technologists involved in using ionizing radiation for diagnostic procedures with children and adolescents, and should also be of value to medical physicists and regulators. It focuses on the measures necessary to provide protection from the effects of radiation using the principles established in the IAEA's Radiation Protection and Safety of Radiation Sources: International Basic Safety Standards (IAEA Safety Standards Series No. GSR Part 3 and the priority accorded to the area). The emphasis throughout is on the special requirements of paediatrics.
3. *ICRP Publication 135 (ICRP 2017). Diagnostic Reference Levels in Medical Imaging.* The publication is intended as a further source of information and guidance on DRLs, recommends quantities for use as DRLs for various imaging modalities and has a dedicated section on paediatric imaging. It suggests modifications in the conduct of DRL surveys that take advantage of automated reporting of radiation dose related quantities and highlights the importance of including information on DRLs in training programmes for healthcare workers.
4. *ICRP Publication 121 (ICRP 2013). Radiological Protection in Paediatric Diagnostic and Interventional Radiology.* This publication aims to provide guiding principles of radiological protection for referring clinicians and clinical staff performing diagnostic imaging and interventional procedures for paediatric patients. It begins with a brief description of the basic concepts of radiological protection, followed by the general aspects of radiological protection, including principles of justification and optimization. It also provides a table with DRL values from various trials and publication.
5. *EC report 185 (EC 2018). European Guidelines on Diagnostic Reference Levels for Paediatric Imaging.* This is the most recent report related to paediatric DRLs. It provides basic recommendations on how to establish and use DRLs for paediatric X-ray examinations and procedures. It also contains tables with national DRL values for paediatric examinations and

procedures in European countries, as well as DRL data from selected publications in European Countries. Finally, the document promotes their use to advance the optimization of radiation protection of paediatric patients, with a focus on CT, interventional procedures using fluoroscopy and digital radiographic imaging. The need to define DRLs separately for different indications if these require different IQ is underlined within the report.

6. *World Health Organization (WHO) report (WHO 2016). Communicating Radiation Risks in Paediatric Imaging.* This document serves as a communication tool about known or potential radiation risks associated with paediatric imaging procedures. It provides information and resources to support communication strategies and has simple tables with typical effective doses for common paediatric diagnostic imaging examinations and their equivalence in terms of number of chest X-rays and duration of exposure to natural background radiation.
7. *AAPM Report 204 (AAPM 2011). Size-Specific Dose Estimates (SSDE) in Paediatric and Adult Body CT Examinations.* The AAPM Task group created conversion factors for providing paediatric patient dose from CTDIw and CTDIv displayed factors taking into consideration patient size.

4.2.1 Paediatric doses and radiation-induced risks associated with radiography

There are some recent publications that discuss the issue of paediatric patient radiation dose measurement in radiography. Some studies use also normalization factors derived from weight and height data. Authors have several options when characterizing paediatric patients. Specific paediatric ages are definitely the most commonly used patient characterization; however, there is a missing link regarding the understanding of what authors are referring to when defining a: 0, 5, 10, 15-year-old paediatric patient. There is a need to build a methodology to define a standard patient that could serve the purpose of harmonizing data collection and assessing paediatric dose.

Dose parameters

The quantities applied in the literature to characterize paediatric dose were mainly:

- Entrance surface dose (ESD);
- Entrance surface air kerma (ESAK);
- Kerma area product (KAP) (or dose-area product, DAP);
- Organ and effective dose (*E*).

The values were either calculated or measured.

Examination technique

These are the most critical factors that influence patient dose exposure. Most of the studies report that the use of anti-scatter grids and additional filtration in paediatric examinations should be considered for the calculation of patient radiation dose as they influence the patient dose. However, a literature review reveals a lack of harmonization in the methodology used for data collection not only regarding the use of examination technical factors but also of dose exposure values, patient characterization, influence of different technology and impact on IQ. The authors' results reflect the working process at their own hospital or region, describing partially only some of the factors: tube voltage (kV), tube current (mA), exposure time (ms). Therefore, it is not possible to identify a common line of procedure. Other important data that are also missing in most of the articles are collimation area, use of grid and added filtration; all of them having tremendous impact both on dose and IQ.

X-ray equipment

The influence of technology and exposure values utilized by each vendor creates a lot of confusion amongst health professionals. Most of the papers published do not take into account IQ.

Selected papers from the recent literature

The papers included in this section are presented according to the year of publication. The most recent papers are shown first in the list and are selected as they report interesting findings.

Pedersen et al (2018)

The aim of the study was to validate the reproducibility of 3D reconstructions of the spine using a new reduced micro-dose protocol using an anthropomorphic child phantom undergoing low-dose biplanar radiography. The authors performed an analysis to establish the lowest dose that enabled acceptable visibility of spinal landmarks. The phantom measurements indicated a reduction of approximately 58% in radiation dose. *In vivo* results showed acceptable but inferior intra- and inter-observer reliability. The authors' conclusion stated that the new protocol offers a preliminary screening option and a follow-up tool for children with mild scoliosis yielding extremely low radiation and could replace micro-dose protocol for these patients.

Vassileva and Rehani (2015)

As has already been mentioned, there is a lot of confusion regarding paediatric patient grouping. There are documents that mention age, weight or other parameters when dealing with dose surveys. The present work aimed to suggest a pragmatic approach to achieve reasonable accuracy for performing patient dose surveys in countries with limited resources. The analysis is based on a subset of data collected within the IAEA survey of paediatric CT doses, involving 82 CT facilities from 32 countries in Asia, Europe, Africa and Latin America. Data for 6115 patients were collected, in 34.5% of which data for weight were available. The present study suggests that using four age groups, <1, >1–5, >5–10 and >10–15 years, is realistic and pragmatic for dose surveys in countries with fewer resources and for establishment of DRLs. To ensure relevant accuracy of results, data for >30 patients in a particular age group should be collected if patient weight is not known. If a smaller sample is used, patient weight should be recorded and median weight in the sample should be within 5%–10% from the median weight of the sample for which DRLs were established. Another important point the authors made was that comparison of results from different surveys should always be performed with caution, taking into consideration the method of grouping paediatric patients.

Jones et al (2015)
The authors of this study evaluate the effect of beam quality on the IQ of ankle radiographs of paediatric patients in the age range of 0–1 year whilst maintaining constant effective dose (*E*). Lateral ankle radiographs of an infant foot phantom were taken at a range of tube potentials (40.0–64.5 kVp) with and without 0.1 mm copper (Cu) filtration. Effective dose to the patient was computed for the default exposure parameters using the PC program for X-ray Monte Carlo (PCXMC) version 2.0 Monte Carlo software and was fixed for other beam qualities by modulating the tube current-time product. The authors state that according to the results of their study, a lower beam quality will produce better IQ with no additional dose penalty for infant extremity imaging.

Seidenbusch et al (2014)
Knowledge of organ and effective doses achieved during paediatric X-ray examinations is an important prerequisite for assessment of the radiation burden to the patient. Conversion coefficients for the reconstruction of organ doses in about 40 organs and tissues from measured entrance doses during pelvis and hip joint radiographs of 0, 1, 5, 10, 15 and 30 year-old patients were calculated for the standard sagittal beam projection and the standard focus detector distance of 115 cm. The conversion coefficients can be used for organ dose assessments from entrance doses measured during pelvis and hip joint radiographs of children and young adults with all field settings within the optimal and suboptimal standard field settings.

Damilakis et al (2013)
There are practically no data in the literature on radiation doses and potential risks following paediatric DXA performed on GE Lunar DXA scanners. This study aimed to estimate effective doses and associated cancer risks involved in paediatric examinations performed on a GE Lunar Prodigy scanner. Four physical anthropomorphic phantoms representing newborn, 1, 5, and 10-year-old patients were employed to simulate DXA exposures. The estimated lifetime cancer risks were negligible, that is, 0.02–0.25 per million, depending on the sex, age and type of DXA examination. The authors reported also a formula for the estimation of effective dose from examinations performed on GE Lunar Prodigy scanners installed in other institutions.

4.2.2 Paediatric doses and radiation-induced risks associated with fluoroscopy

A literature review shows that a very limited number of studies focus on the issue of paediatric dose assessment (in the last 10 years published papers are slightly over ten). Most of the papers include data from a single paediatric institution and only one study originated from a multicentric national survey.

Types of exams

- Micturating cystourethrography;
- Barium meal;
- Barium swallow;
- Upper gastrointestinal (UGI) examinations.

Patient grouping

In all but one study, patients were grouped according to age. The grouping was done in a similar way as in radiography. There was one study in which grouping was done according to sex and age.

Dose parameters

- KAP or DAP;
- Effective dose;
- Organ doses.

Selected papers from the recent literature: fluoroscopy

Yakoumakis et al (2015)

The purpose of the particular study was to estimate the organ and effective doses of paediatric patients undergoing barium meal examinations and also to evaluate the assessment of radiation Risk of Exposure Induced cancer Death (REID). This was accomplished by using clinical measurements of DAP and the PCXMC 2.0 Monte Carlo code. For all ages, the main contributors to the total organ and effective doses were the fluoroscopy projections. The average DAP values and absorbed doses to the patient were higher for the left lateral projections. The REID was calculated for boys (4.8×10^{-2}%, 3.0×10^{-2}% and 2.0×10^{-2}%) for neonatal, 1 and 5-year-old patients, respectively. The corresponding values for girl patients were calculated (12.1×10^{-2}%, 5.5×10^{-2}% and 3.4×10^{-2}%).

Yakoumakis et al (2013)

The authors estimated organ and effective doses of paediatric patients undergoing micturating cystourethrography examinations. DAP values of 90 patients undergoing cystourethrography examinations were recorded. These were used together with two Monte Carlo codes, MCNP5 and PCXMC 2.0, to assess the organ doses in these procedures. The organs receiving the highest radiation doses were the urinary bladder (ranging from 1.9 mSv in the newborn to 4.7 mSv in a 5-year-old patient) and the large intestines (ranging from 1.5 mSv in the newborn to 3.1 mSv in the 5-year-old patient). They found reasonable agreement between the dose estimates provided by PCXMC v2.0 and MCNP5 codes for most of the organs. In special cases, though, there were systematic disagreements in organ doses such as in the skeleton, gonads and esophagus.

Emigh et al (2013)
The authors deal with DAP and effective dose estimation for paediatric upper gastrointestinal (UGI) examinations. 649 consecutive UGI studies from a single institution were reviewed to define a standard protocol examination. This standard protocol included 3.6 min of fluoroscopy and 4 spot exposures. Anthropomorphic phantoms with MOSFET dosimeters were used for effective dose estimation in four age groups: newborn, 1, 5 and 10 years old. For the above-mentioned age groups DAP ($Gy \cdot cm^2$) was 0.08, 0.28, 0.82 and 1.22, respectively. Measured effective doses ranged from 0.35 to 0.79 mSv in girls and were 3%–8% lower for boys.

Hsi et al (2013)
This is the only study found on videourodynamics. Fluoroscopy data were collected together with dose metrics in order to calculate entrance skin dose after applying a series of correction factors. Effective doses and organ specific doses (ovaries/testes) were also estimated from entrance skin dose using Monte Carlo methods on a mathematical anthropomorphic phantom (ages 0, 1, 5, 10 and 15 years). On multivariate adjusted analysis, BMI, bladder capacity and fluoroscopy time were independently associated with effective dose. The authors also found that the average effective dose from videourodynamics was less compared to the voiding cystourethrogram dose reported in the literature. Greater fluoroscopy time, BMI and bladder capacity are independently associated with higher dosing.

4.2.3 Paediatric doses and radiation-induced risks associated with CT

A literature review results in more than 700 papers on CT paediatric radiation dose and more than 100 papers on CT paediatric effective dose and CT paediatric organ dose. In the last few years, the literature consists of publications dealing with the issue of radiation-induced risk from CT based mainly on epidemiological data and not on strict patient dose measurements or calculations.

Patient grouping
The most common groupings found in the literature are based on:
- Age: <1, 1–5, 5–10, 10–15 years;
- Weight;
- Age and weight;
- Patient width.

Dose parameters
- CTDIw/CTDIv;
- DLP;
- Organ dose;

- Effective dose;
- SSDE.

Exam type
The most common examinations for which patient dose was reported were:
- Brain;
- Chest;
- Abdomen.

There are also publications focused on:
- Sinuses;
- High resolution chest;
- Temporal bone;
- Lumbar spine;
- Low dose chest.

Selected papers from the recent literature
Some interesting publications in the recent literature are described in more detail in this section.

Meulepas et al (2018)
The authors studied a nationwide retrospective cohort of 168 394 children who received one or more CT scans in a Dutch hospital between 1979 and 2012 who were younger than 18 years of age. They obtained cancer incidence, vital status and confounder information by record linkage with external registries. They calculated standardized incidence ratios using cancer incidence rates from the general Dutch population. Excess relative risks were also calculated. They found evidence that CT-related radiation exposure increases brain tumor risk. No association was observed for leukemia. Compared with the general population, the incidence of brain tumors was higher in the cohort of children with CT scans, requiring cautious interpretation of the findings.

Habib Geryes et al (2018)
The authors proved that the use of step-and-shoot mode in paediatric cardiac CT angiography is possible at heart rates greater than 65 bpm, allowing low-dose acquisition with single-source 64-slices CT. More specifically, for coronary indications the median effective dose was reduced from 0.9 to 0.6 mSv. For whole thorax indications, corresponding values were 1.1 mSv from initial values of 2.7 mSv.

Kobayashi et al (2018)
The purpose of this study was to examine intraoperative radiation exposure in paediatric spinal scoliosis surgery using O-arm. This is a navigation system that allows intraoperative imaging that facilitates highly accurate instrumentation surgery. Using O-arm, the median dose per scan was 92.5 mGy, and mean total dose was 401 mGy. The authors stated that this amount of radiation dose was more

than an 80% reduction of mean preoperative CT dose of 460 mGy and thus the use of O-arm in the original protocol contributed to a reduction in radiation exposure.

Inkoom et al (2015)
The aim of this study was to determine the location of radiosensitive organs in the interior of four paediatric anthropomorphic phantoms for dosimetric purposes. The authors used four paediatric anthropomorphic phantoms representing the average individual as a newborn, 1, 5 and 10-year-old child who underwent head, thorax and abdomen CT scans. CT scans of all children aged 0–16 years performed in 5 years in the hospital were reviewed. 503 were found to be eligible for normal anatomy. According to the authors, the production of charts of radiosensitive organs inside paediatric anthropomorphic phantoms was feasible and may provide users with reliable data for positioning of dosimeters during direct organ dose measurements.

Mathews et al (2013)
They authors aimed to assess the cancer risk in children and adolescents following exposure to low-dose ionizing radiation from diagnostic CT scans. They identified 10.9 million people from Australian Medicare records, aged 0–19 years on 1 January 1985 or born between 1 January 1985 and 31 December 2005; all exposures to CT scans funded by Medicare during 1985–2005 were identified for this cohort. Cancers diagnosed in cohort members up to 31 December 2007 were obtained through linkage to national cancer records. They found that an increased incidence of cancer after CT scan exposure in this cohort was mostly due to irradiation.

4.2.4 Paediatric doses and radiation-induced risks associated with interventional procedures

A literature review reveals a rather large number of interventional cardiology compared to radiology procedures.

Types of exams

- Balloon valvuloplasty;
- Angioplasty;
- Patent ductus arteriosus (PDA) occlusion;
- Atrial septal defect (ASD) occlusion;
- Electrophysiology;
- Aortic angioplasty;
- Pulmonary angioplasty with stent;
- Atrial septal defect closure;
- Aortic valvuloplasty;
- Pulmonary valvuloplasty;
- PDA closure with coil;
- PDA closure with device;
- And many other procedures.

Patient grouping
Patients were grouped mainly according to age or weight or both.

Dose parameters
- KAP (DAP);
- Air kerma;
- Effective dose;
- Organ doses.

Selected papers from the recent literature: interventional cardiology
Karambatsakidou et al (2019)
The authors estimated *E*, equivalent organ doses *H* and associated conversion coefficients in paediatric cardiac interventions, using detailed exposure data from radiation dose structured reports. These are also compared with estimations performed using the approach currently implemented in the clinic that is based on a simplified assumptions method. The work was performed using the Monte Carlo system PCXMC to calculate *E* and equivalent organ doses for 202 children. The calculations were performed with input values from the Radiation Dose Structured Reports (RDSR), and also using simplified assumptions, including fixed nominal values for the focus-skin distance, collimated beam size, irradiation geometry and patient size (age, weight and height).

Peters et al (2015)
Data on effective dose (*E*) in 3D rotational angiography in children are lacking. The purpose of this study was to provide *E* of 3D rotational angiography (RA) and to correlate this with parameters readily available in daily practice. Effective doses were calculated with Monte Carlo PCXMC 2.0 in 14 patients who underwent a total of 17 3D RAs in a single paediatric catheterization laboratory. The median age was 5.7 years (range 1 day–16.6 years). Median *E* was 1.6 mSv (range 0.7–4.9). *E* did not correlate with age and body surface area but did correlate with DAP and mGy with *r*(2) of 0.75 and 0.83, respectively. Reduction of the total amount of frames from 248 to 133 per rotation resulted in further dose reduction of over 50% with preserved IQ.

Kawasaki et al (2015)
This is an interesting study looking over the radiation exposure to neonates and infants during cardiac catheterizations. It is known that smaller patient size and higher heart rate result in a greater need for magnification modes and higher frame rates, all of which contribute to a significant increase in radiation doses. The authors reported their work on evaluating organ and effective doses for neonates and infants during diagnostic cardiac catheterizations, on the basis of in-phantom dosimetry and conversion factors from DAP to the effective dose. Organ doses for 0 and 1-year-old children during diagnostic cardiac catheterizations were measured by radiophotoluminescence glass dosimeters implanted in neonate and infant anthropomorphic phantoms. The mean effective doses evaluated according to ICRP 103 were 7.7 mSv (range, 0.1–18.4 mSv) for a neonate and 7.3 mSv (range 1.9–18.6 mSv)

for an infant. The authors provided conversion factors from DAP to the effective dose. These were 2.2 and 4.0 in posteroanterior and lateral cine angiography, respectively, for a neonate and 1.4 and 2.7 in posteroanterior and lateral cine angiography, respectively, for an infant.

Kobayashi et al (2014)
Kobayasji *et al* performed a multicenter observational study of radiation dose in paediatric laboratories. Patient demographic, procedural and radiation data including fluoroscopic time and dose area product (DSA) in μGy m^2 were analyzed. PKA/body weight (BW) was obtained by indexing DSA to BW. A total of 8267 paediatric catheterization procedures (age <18 years) were included from 16 institutions. The procedures consisted of diagnostic (n = 2827), transplant right ventricular (RV) biopsy (n = 1172), and interventional catheterizations (n = 4268). DSA correlated with BW better than with age and best correlated with weight-fluoroscopic time product. DSA/BW showed consistent values across paediatric ages. Interventional catheterizations had the highest DSA/BW (50th, 75th and 90th percentiles: 72, 151 and 281 μGy · m^2 kg^{-1}), followed by diagnostic (59, 105 and 175 μGm^2 kg^{-1}) and transplant RV biopsy (27, 79 and 114 μGy · m^2 kg^{-1}). According to the authors, DSA/BW appeared to be the most reliable standard to report radiation dose across all procedure types and patient age. Therefore, they recommended DSA/BW to be used as the standard unit in documenting radiation usage in paediatric laboratories.

Barnaoui et al (2014)
The paper deals with local DRLs, effective dose and organ doses from interventional cardiology procedures in a French paediatric reference center. 801 procedures performed between 2010 and 2011 were considered. Dosimetric data including DAP, fluoroscopy time (FT) and number of cine frames (NF) were analyzed. Patients were categorized according to weight. Doses to the lungs, esophagus, breast and thyroid were evaluated using anthropomorphic phantoms and thermoluminescent dosimeters. Effective doses were calculated using DAP and conversion factors calculated with PCXMC 2.0 software. Beside diagnostic procedures, five therapeutic procedures were considered: balloon valvuloplasty, angioplasty, PDA occlusion, ASD occlusion and electrophysiology. Patients were categorized according to BW: ⩽6.5 kg, 6.5–14 kg, 14.5–25.5 kg, 25.5–43.5 kg, >43.5 kg. Large variations of dose exposure were observed for a given procedure and given BW group. Organ exposure was high in the lungs and esophagus, especially in newborns.

References

AAPM 2011 *Size-Specific Dose Estimates (SSDE) in Pediatric and Adult Body CT Examinations* Report of the 204 Task group AAPM

Barnaoui S, Rehel J L, Baysson H, Boudjemline Y, Girodon B, Bernier M O, Bonnet D and Aubert B 2014 Local reference levels and organ doses from paediatric cardiac interventional procedures *Pediatr. Cardiol.* **35** 1037–45

Damilakis J, Perisinakis K, Grammatikakis J, Panayiotakis G and Gourtsoyiannis N 1996 Accidental embryo irradiation during barium enema examinations. An estimation of absorbed dose *Invest. Radiol* **31** 242–5

Damilakis J, Perisinakis K, Voloudaki A and Gourtsoyiannis N 2000 Estimation of fetal radiation dose from computed tomography scanning in late pregnancy: depth-dose data from routine examinations *Invest. Radiol.* **35** 527–33

Damilakis J, Perisinakis K, Vrahoriti H and Kontakis G *et al* 2002 Embryo/fetus radiation dose and risk from dual x-ray absorptiometry examinations *Osteoporos. Int.* **13** 716–22

Damilakis J, Perisinakis K, Prassopoulos P, Dimovasili E, Varveris H and Gourtsoyiannis N 2003a Conceptus radiation dose and risk from chest screen-film radiography *Eur. Radiol.* **13** 406–12

Damilakis J, Solomou G, Manios G E and Karantanas A 2013 Paediatric radiation dose and risk from bone density measurements using a GE lunar prodigy scanner *Osteoporos. Int.* **24** 2025–31

Damilakis J, Theocharopoulos N, Perisinakis K, Manios E, Dimitriou P, Vardas P and Gourtsoyiannis N 2001 Conceptus radiation dose and risk from cardiac catheter ablation procedures *Circulation* **104** 893–7

Damilakis J, Theocharopoulos N, Perisinakis K, Papadokostakis G, Hadjipavlou A and Gourtsoyiannis N 2003b Conceptus radiation dose assessment from fluoroscopically assisted surgical treatment of hip fractures *Med. Phys.* **30** 2594–601

EC Radiation Protection report 185 2018 European Guidelines on Diagnostic Reference Levels for Paediatric Imaging Directorate-General for Energy, Directorate D — Nuclear Energy, Safety and ITER, Unit D3 — Radiation Protection and Nuclear Safety

Emigh B, Gordon C L, Connolly B L, Falkiner M and Thomas K E 2013 Effective dose estimation for paediatric upper gastrointestinal examinations using an anthropomorphic phantom set and metal oxide semiconductor field-effect transistor (MOSFET) technology *Pediatr. Radiol.* **43** 1108–16

Gu J, Xu X, Caracappa P and Liu B 2013 Fetal doses to pregnant patients from CT with tube current modulation calculated using Monte Carlo simulations and realistic phantoms *Rad. Prot. Dosim.* **155** 64–72

Habib Geryes B, Calmon R, Donciu V, Khraiche D, Warin-Fresse K, Bonnet D, Boddaert N and Raimondi F 2018 Low-dose paediatric cardiac and thoracic computed tomography with prospective triggering: is it possible at any heart rate? *Phys. Med.* **49** 99–104

Hsi R S, Dearn J, Dean M, Zamora D A, Kanal K M, Harper J D and Merguerian P A 2013 Effective and organ specific radiation doses from videourodynamics in children *J. Urol.* **190** 1364–9

IAEA 2014 Dosimetry in Diagnostic Radiology for Paediatric Patients *Human Health Series* 24 (Vienna: IAEA)

IAEA 2013 Radiation Protection in Paediatric Radiology *Safety Reports Series* 71 (Vienna: IAEA)

ICRP 2013 Radiological protection in paediatric diagnostic and interventional radiology. ICRP Publication 121

ICRP 2017 Publication 135, Diagnostic reference levels in medical imaging

ICRP Publication 2000 Pregnancy and Medical Radiation *Safety Reports Series* 84 (Vienna: IAEA)

Inkoom S, Raissaki M, Perisinakis K, Maris T G and Damilakis J 2015 Location of radiosensitive organs inside pediatric anthropomorphic phantoms: data required for dosimetry *Phys. Med.* **31** 882–8

Jones A, Ansell C, Jerrom C and Honey I D 2015 Optimization of image quality and patient dose in radiographs of paediatric extremities using direct digital radiography *Br. J. Radiol.* **88** 20140660

Karambatsakidou A, Omar A, Fransson A and Poludniowski G 2019 Calculating organ and effective doses in paediatric interventional cardiac radiology based on DICOM structured reports – is detailed examination data critical to dose estimates? *Phys. Med. Jan.* **57** 17–24

Kawasaki T, Fujii K and Akahane K 2015 Estimation of organ and effective doses for neonate and infant diagnostic cardiac catheterizations *Am. J. Roentgenol.* **205** 599–603

Kelaranta A, Makela T, Kaasalainen T and Kortesniemi M 2017 Fetal radiation dose in three common CT examinations during pregnancy—Monte Carlo study *Phys. Med.* **43** 199–206

Kobayashi K, Ando K, Ito K, Tsushima M, Morozumi M, Tanaka S, Machino M, Ota K, Ishiguro N and Imagama S 2018 Intraoperative radiation exposure in spinal scoliosis surgery for pediatric patients using the O-arm® imaging system *Eur. J. Orthop. Surg. Traumatol.* **28** 579–83

Kobayashi D *et al* 2014 Standardizing radiation dose reporting in the pediatric cardiac catheterization laboratory-a multicenter study by the CCISC (congenital cardiovascular interventional study consortium) *Catheter. Cardiovasc. Interv.* **84** 785–93

Mathews J D *et al* 2013 Cancer risk in 680 000 people exposed to CT scans in childhood or adolescence: data linkage study of 11 million Australians *Brit. Med. J.* **21** f2360

Metzger R L and Van Riper K 1999 Fetal dose assessment from invasive special procedures by Monte Carlo methods *Med. Phys.* **26** 1714–20

Meulepas J M *et al* 2019 Radiation Exposure from paediatric CT scans and subsequent cancer risk in the Netherlands *J. Natl. Cancer Inst.* **111** 256–63

National Council on Radiation Protection and Measurements (NCRP) 1994 *Considerations Regarding the Unintended Radiation Exposure of the Embryo, Fetus or Nursing Child* NCRP commentary no. 9 (Bethesda, Maryland: NCRP Publications), pp 11–2

Osei E and Faulkner K 1999 Fetal doses from radiological examinations *Brit. J. Radiol.* **72** 773–80

Pedersen P H, Vergari C, Alzakri A, Vialle R and Skalli W 2019 A reduced micro-dose protocol for 3D reconstruction of the spine in children with scoliosis: results of a phantom-based and clinically validated study using stereo-radiography *Eur. Radiol.* **29** 1874–81

Perisinakis K, Seimenis I, Tzedakis A and Damilakis J 2014 Perfusion scintigraphy versus 256-slice CT angiography in pregnant patients suspected of pulmonary embolism: comparison of radiation risks *J. Nucl. Med.* **55** 1–8

Peters M, Krings G, Koster M, Molenschot M, Freund M W and Breur J M 2015 Effective radiation dosage of three-dimensional rotational angiography in children *Europace* **17** 611–6

Samara E, Stratakis J, Enele Melono J M, Mouzas I A, Perisinakis K and Damilakis J 2009 Therapeutic ERCP and pregnancy: is the radiation risk for the embryo trivial? *Gastrointest. Endosc.* **69** 824–31

Seidenbusch M C and Schneider K 2014 Conversion coefficients for determining organ doses in paediatric pelvis and hip joint radiography *Pediatr. Radiol.* **44** 1110–23

Stabin M G, Watson E E, Cristy M, Ryman J C, Eckerman K F, Davis J L, Marshall D and Gehlen M K 1995 *Mathematical Models and Specific Absorbed Fractions of Photon Energy in the Nonpregnant Adult Female and at the End of Each Trimester of Pregnancy* ORNL/TM-1907 Oak Ridge National Laboratory Report

Theocharopoulos N, Damilakis J, Perisinakis K, Papadokostakis G, Hadjipavlou A and Gourtsoyiannis N 2006 Fluoroscopically assisted surgical treatments of spinal disorders: Conceptus radiation doses and risks *Spine* **2** 239–44

Vassileva J and Rehani M 2015 Patient grouping for dose surveys and establishment of diagnostic reference levels in paediatric computed tomography *Radiat. Prot. Dosim.* **165** 81–5

White W M, Zite N B, Gash J, Waters W B, Thompson W and Klein F A 2007 Low-dose computed tomography for the evaluation of flank pain in the pregnant population *J. Endourol.* **21** 1255–60

WHO report 2016 Communicating Radiation Risks in Paediatric Imaging

Xu G, Taranenko V, Zhang J and Shi C 2007 A boundary-representation method for designing a whole-body radiation dosimetry models: pregnant females at the ends of three gestational periods—RPI-P3, -P6 and -P9 *Phys. Med. Biol.* **52** 7023–44

Yakoumakis E, Dimitriadis A, Gialousis G, Makri T, Karavasilis E, Yakoumakis N and Georgiou E 2015 Evaluation of organ and effective doses during paediatric barium meal examinations using PCXMC 2.0 Monte Carlo code *Radiat. Prot. Dosim.* **163** 202–9

Yakoumakis E, Dimitriadis A, Makri T, Karlatira M, Karavasilis E and Gialousis G 2013 Verification of radiation dose calculations during paediatric cystourethrography examinations using MCNP5 and PCXMC 2.0 Monte Carlo codes *Radiat. Prot. Dosim.* **157** 355–62

Chapter 5

Methods to calculate conceptus and paediatric dose

J Damilakis and V Tsapaki

Radiation doses from the same type of X-ray examination may vary considerably depending on many factors such as the equipment used, the exposure parameters selected by the operator and use of dose reduction tools during the procedure. Therefore, dose values from the literature provide only general information about dose levels from an examination. Medical physicists need to know how to estimate doses from X-ray imaging accurately. Methods to calculate conceptus dose from diagnostic and interventional procedures as well as methods to calculate paediatric doses from various types of X-ray examinations are described below.

5.1 Methods to calculate conceptus dose from diagnostic and interventional procedures

A substantial increase in the use of radiological procedures for pregnant patients has been observed over the last few decades, while accidental medical exposure of pregnant patients during the first post-conception weeks may occasionally lead to unnecessary pregnancy terminations. Information on dose estimation is essential for the justification of the exposure as well as for dose optimization. Moreover, conceptus dose information is needed for risk communication. The expectant mother needs to know the level of radiation burden to the conceptus from in-utero exposure before the examination. Therefore, determination of the conceptus dose to pregnant patients undergoing radiological procedures is of paramount importance for the dose management of the expectant mother.

Assessment of the conceptus dose from radiological examinations is possible in everyday clinical practice by using methods published in the literature or software packages such as the Conceptus Dose Estimation (CoDE) web-based tool. These

doi:10.1088/978-0-7503-1317-9ch5

methods are capable of estimating conceptus dose during the first post-conception weeks and at the end of the first, second and third trimesters of gestation. Sometimes uterus dose is used as a surrogate of conceptus dose. Estimation of uterus dose to a patient who has undergone an X-ray examination can be made using methods based on dose indices such as dose-area product (DAP) or dose length product (DLP) and conversion coefficients. In the absence of a more specific method, these methods can used in clinical routine for conceptus dose estimation during the first weeks of gestation provided that the patient's body size is similar to that of the phantom used in the method.

5.1.1 Methods published in the literature to calculate conceptus dose from radiographic and fluoroscopic examinations

Damilakis *et al* (2002)

This work provides conceptus doses normalized to air kerma for anteroposterior (AP) and posteroanterior (PA) abdominal radiographic and fluoroscopic exposures that can be used for conceptus dose estimation for all trimesters of gestation.

Conceptus dose calculation is possible using the equation

$$D_c = D_n K_{\text{air}}$$

where D_c is the dose to the conceptus, D_n is the normalized conceptus dose value, the K_{air} is the air kerma.

Data provided in the above publication apply for focus-to-surface distance (FSD) equal to 100 cm for both AP and PA projections for all trimesters of gestation. For different FSD, conceptus dose can be calculated using the following formula

$$D_c = D_n(K_{\text{air}})\left(\frac{(\text{SED}' - T')(\text{SED} - T + d)}{(\text{SED} - T)(\text{SED}' - T' + d)}\right)^2$$

where D_c is the dose to the conceptus, D_n is the normalized conceptus dose value, K_{air} is the air kerma, SED′ is the skin to exit distance of interest, T' is the AP thickness of the patient, SED is the skin to exit distance of the data provided in tables that is 120 cm for the first trimester, 126.4 cm for the second trimester and 129.4 cm for the third trimester, T is the AP thickness of the phantom that is 20 cm for the first trimester, 26.4 cm for the second trimester and 29.4 cm for the third trimester, and d is the depth of the conceptus from the incidence surface.

Damilakis *et al* (2001)

The majority of patients who need cardiac catheter ablation procedures are young and some of them are female patients of childbearing age. The scientific literature shows that these fluoroscopically guided examinations are associated with high radiation doses. Cardiac ablation procedures are rarely performed on pregnant patients. However, there are cases of accidental exposure of pregnant patients. A method of estimating conceptus dose during all trimesters of gestation has been developed by Damilakis *et al* (2001).

At the beginning of the cardiac ablation procedure, the electrode catheters are inserted and advanced from the groin to the heart using fluoroscopy in the posteroanterior projection (GHPA projection). Afterward, the catheters are positioned under fluoroscopic control for PA, left anterior oblique (LAO 45°), and right anterior oblique (RAO 25°) projections of the heart. The conceptus dose D_c at the gestational period j ($j = 1$ for the first, 2 for the second and 3 for the third trimester) for a cardiac ablation procedure performed in an X-ray unit S may be calculated by using the formula

$$D_c^j = \sum_{i=1}^{4}\left[\left(\frac{O_S}{O}\right)_i \cdot \left(\frac{\text{mAs}}{\text{mA}}\right)_i \cdot (t_S)_i \cdot \left(\frac{\text{SSD}}{\text{SSDs}}\right)_i^2 \cdot D_i^j\right]$$

where i is the projection taken during the cardiac ablation procedure ($i = 1, 2, 3$, and 4 for the GHPA, PA, RAO, and LAO projections of the heart, respectively); O_S is the output of the system S at 70 cm, estimated at the mean kVp set for the projection i; mAs, t_S, and SSDs are the tube current, fluoroscopy time, and source-to-surface distance, respectively, used for the projection i taken in the X-ray system S; O is the output of the X-ray tube of the system used in the study published by Damilakis *et al* (2001) estimated at the kVp set for projection i; mA and SSD are the tube current and source-to-surface distance, respectively, used for projection i taken in the system used in the study (Damilakis *et al* 2001); and D_i^j is the dose per minute of fluoroscopy for the projection i estimated in the study published by Damilakis *et al* (2001). Values of doses per minute of fluoroscopy measured during all trimesters of gestation can be found in the published study.

Samara *et al* (2009)

Choledocholithiasis can occur in a considerable percentage of the pregnant population (up to 12%) and increases with gestational age. Medical intervention cannot be postponed because of the likelihood of cholangitis and pancreatitis. Endoscopic retrograde cholangiopancreatography (ERCP) is among the main non-surgical procedures that can be performed for the treatment of choledochal stones and gallstones pancreatitis. One of the main limitations for applying ERCP is the radiation dose absorbed by the unborn child and the related radiogenic risks. Conceptus dose estimation is possible by multiplying the DAP value of the ERCP procedure with DAP-normalized data (D_c^{norm}) provided in the work by Samara *et al* (2009):

$$D_c = D_c^{\text{norm}} \cdot \text{DAP}$$

Normalized conceptus dose data are presented as a function of kVp, total filtration, maternal body mass index and gestational stage.

Solomou *et al* (2016),

Placenta accreta is a serious pregnancy condition that occurs when the placenta attaches abnormally to the uterine wall. This can cause severe bleeding after delivery. Treatment may be delivery by caesarian section followed by hysterectomy that is surgical removal of the uterus. The prophylactic hypogastric artery balloon occlusion (HABO) with or without adjunct embolization under fluoroscopic guidance is a recognized alternative treatment in parturient women for the management of obstetric hemorrhage. The fluoroscopic projections commonly used in HABO are the PA, the RAO and the LAO.

A study (Solomou *et al* 2016) has produced projection-specific normalized conceptus dose data, which allow for the accurate estimation of fetal dose in pregnant patients undergoing HABO procedures. During Monte Carlo simulation experiments, the center of the X-ray field was set on the area of the left internal iliac artery of the mathematical phantom (reference field location). The reference X-ray field size was 18 × 22.5 cm (figure 5.1). To study the effect of the location of the X-ray field relative to the fetus on normalised fetal dose (NFD), additional measurements were obtained changing the position of the center of the three projections along both *x*- and *z*-axes. Figure 5.2 illustrates the four extreme locations of the AP field.

The conceptus dose (D_c) from a HABO procedure performed on a parturient woman may be estimated from

$$D_c = \sum_{\text{projection}} (\text{DAP}_{\text{projection}} \cdot \text{NFD}_{\text{projection}} \cdot \text{CF}_{\text{FL}} \cdot \text{CF}_{\text{FS}} \cdot \text{CF}_{\text{PS}})$$

where projection indicates PA, LAO 20° and RAO 20° fluoroscopic projection centered on the left and right iliac artery, $\text{DAP}_{\text{projection}}$ and NFD projection are the cumulative DAP and NFD values for the specific projection, respectively, CF_{FL} and

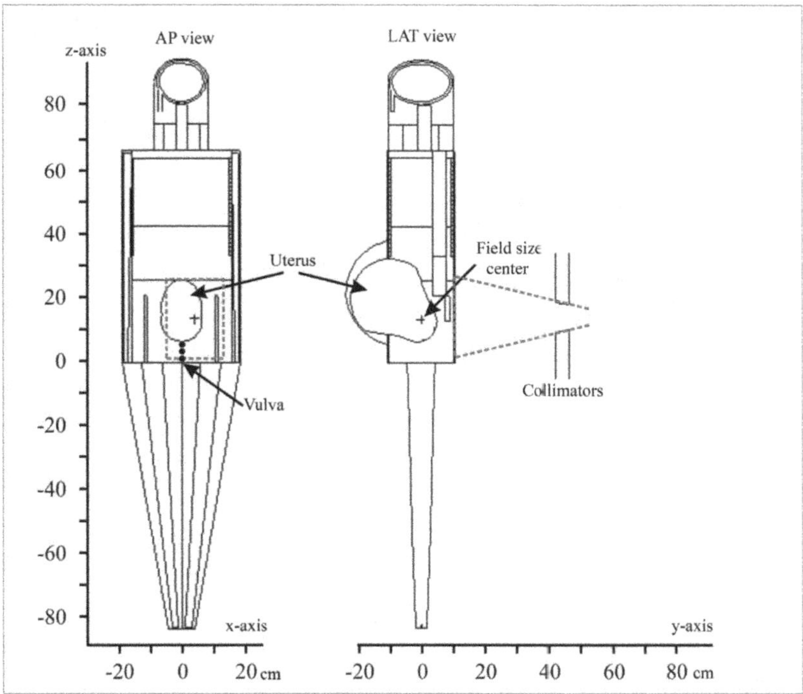

Figure 5.1. An anteroposterior (a) and lateral (b) view of the mathematical phantom used by Solomou *et al.* (2016) The origin of the axes was assumed at the scrotum. The red square indicates the reference X-ray field. The black cross illustrates the center of the reference field assumed in the Monte Carlo simulation of the left internal iliac artery fluoroscopic imaging. Black dots between the uterus and scrotum denote dose measuring points in the vaginal fornix of the phantom.

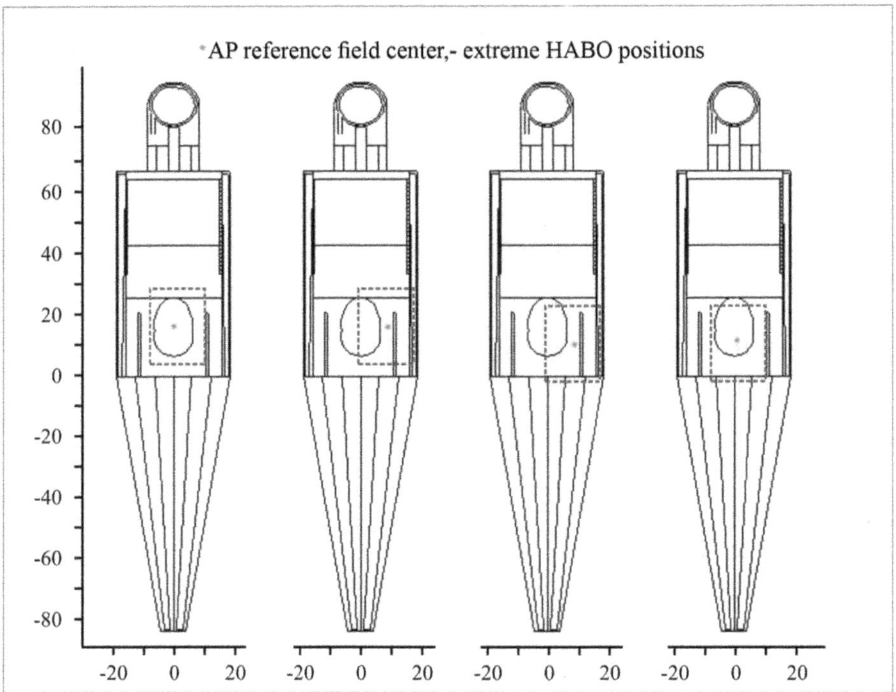

Figure 5.2. The four extreme positions of the AP reference field are illustrated. Eighteen different positions of the reference AP X-ray field center were assumed with x and z coordinates changed by steps of 3 and 1 cm, respectively.

CF_{FS} are the correction factors for X-ray field location and size, respectively, and CF_{PS} is the correction factor applied if a pregnant body size differs from the average body size of the phantom used in the study. The abdominal perimeter of the phantom at the average body size was 108.2 cm.

DAP-normalized fetal dose values for the PA, LAO 20° and RAO 20° fluoroscopic projections performed during the catheterization of the left internal iliac artery and figures presenting the effect of (a) the location of the X-ray field center, (b) the X-ray field size at entrance skin surface and (c) maternal body size on the conceptus radiation dose can be found in the article published by Solomou *et al* (2016).

5.1.2 Methods published in the literature to calculate conceptus dose from CT

Damilakis *et al* (2010)

A general-purpose Monte Carlo simulation code Monte-Carlo-N-particle (MCNP) and four mathematical phantoms simulating pregnancy during all trimesters of gestation (figure 5.3) were used to provide normalized dose data for conceptus dose estimation from any CT examination performed in the trunk of the pregnant patient. The Siemens Sensation 16 and Sensation 64 CT scanners were modeled. The contribution to the conceptus dose from single scans was obtained at various positions across the phantoms. Dose values normalized to CTDI free-in-air for single slices

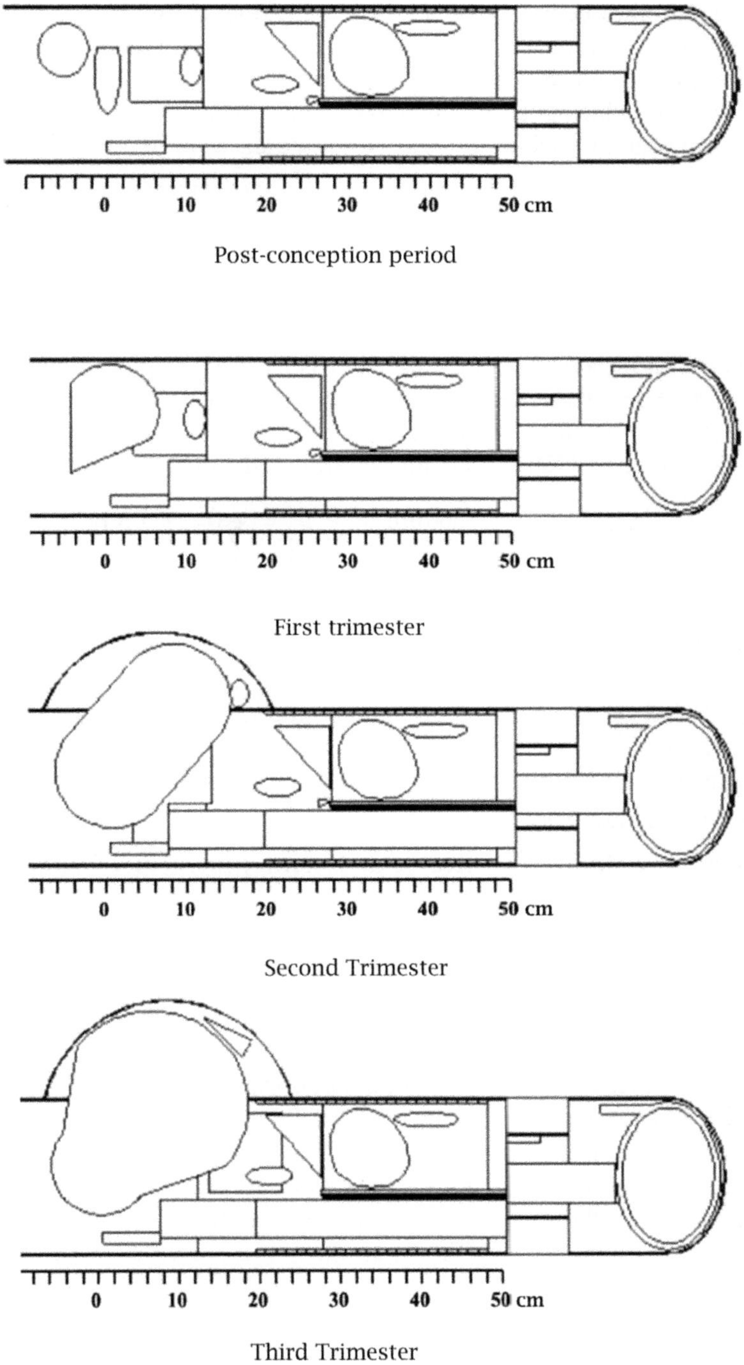

Figure 5.3. The four mathematical phantoms simulating pregnancy used to provide normalized dose data for conceptus dose estimation from CT examinations.

across the phantom representing (a) a female individual of childbearing age and (b) a pregnant individual during the first, second and third trimesters of gestation can be found in the article published by Damilakis *et al* (2010).

Conceptus dose during early pregnancy (i.e. during the first post-conception weeks or during the first trimester) can be estimated using the following equation:

$$D(p_x, d_x) = \mathrm{CTDI}_F \cdot \mathrm{NCD}(p_0, d_0) \cdot \left(\frac{\mathrm{NCD}_{d_x}}{\mathrm{NCD}_{d9}}\right)_{p_x}$$

where CTDI_F is the CTDI free-in-air value measured at the same kVp and mAs selected for the patient examination, $\mathrm{NCD}(p_0, d_0)$ is the normalized conceptus dose for a patient with an average body size that is $p_0 = 88.7$ cm and conceptus depth $d_0 = 9$ cm, NCD_{d_x} is the normalized conceptus dose at depth d_x and NCD_{d9} is the normalized conceptus dose calculated for the phantom simulating the average patient at $d_0 = 9$ cm. $\mathrm{NCD}(p_0, d_0)$ can be estimated using the equation

$$\mathrm{NCD}(p_0, d_0) = \frac{20\ \mathrm{mm}}{\mathrm{BC(mm)}} \cdot \frac{1}{\mathrm{pitch}} \cdot \sum_{z1}^{z2} f_z \left(\frac{\mathrm{mGy}}{\mathrm{mGy_{air}}}\right)$$

where BC is the beam collimation in mm, $z1$ and $z2$ the boundaries of the scanned volume, and f_z the normalized conceptus dose coefficient from a single scan defined as the conceptus dose from a single scan located z cm away from the conceptus along the z-axis divided by the CTDI free-in-air value measured at the same scanning parameters (dose values for the first post-conception weeks and the first trimester of gestation can be found in the article published by Damilakis *et al* 2010). The ratio $\frac{\mathrm{NCD}_{d_x}}{\mathrm{NCD}_{d9}}$ can be obtained from figure 5.4.

Conceptus dose during the second and third trimesters of gestation can be estimated using the following equation:

$$D(p_x, d_x) = \mathrm{CTDI}_F \cdot \mathrm{NCD}(p_0) \cdot \left(\frac{\mathrm{NCD}}{\mathrm{NCD}_0}\right)_{p_x}$$

where CTDI_F is the CTDI free-in-air value measured at the same kVp and mAs selected for the patient examination, $\mathrm{NCD}(p_0)$ is the normalized conceptus dose for a patient with an average body size that is $p_0 = 102.3$ cm for the second trimester and $p_0 = 108.2$ cm for the third trimester, NCD is the normalized conceptus dose at the second or third trimester and NCD_0 is the normalized conceptus dose calculated for the phantom simulating the average patient at the second or third trimester. $\mathrm{NCD}(p_0)$ can be estimated using the equation

$$\mathrm{NCD}(p_0) = \frac{20\ \mathrm{mm}}{\mathrm{BC(mm)}} \cdot \frac{1}{\mathrm{pitch}} \cdot \sum_{z1}^{z2} f_z \left(\frac{\mathrm{mGy}}{\mathrm{mGy_{air}}}\right)$$

where BC is the beam collimation in mm, $z1$ and $z2$ the boundaries of the scanned volume and f_z the normalized conceptus dose coefficient from a single scan defined as the conceptus dose from a single scan located z cm away from the conceptus along

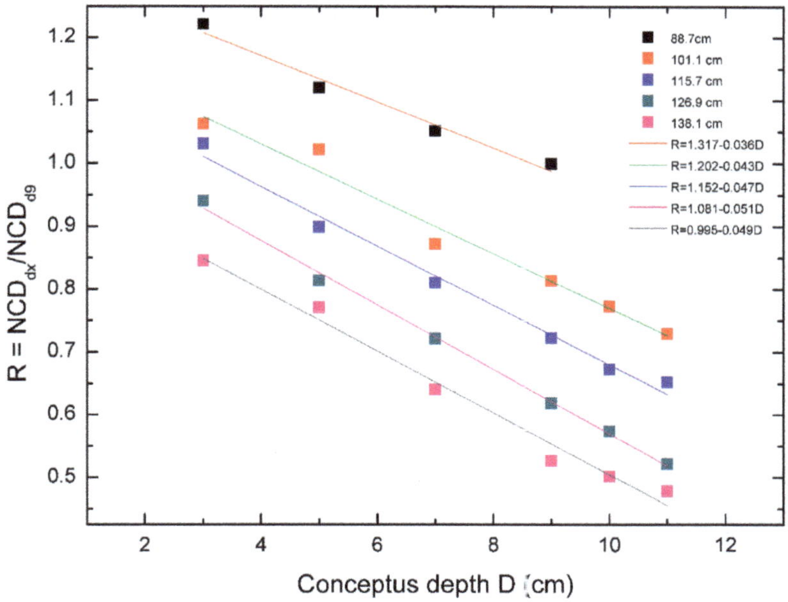

Figure 5.4. The linear relationship between $\frac{\mathrm{NCD}_{d_x}}{\mathrm{NCD}_9}$ and surface-to-conceptus distance for the phantoms of different sizes representing pregnancy during the first post-conception weeks. The perimeters of the slice containing the central area of the conceptus were 88.7, 101.1, 115.7, 126.9 and 138.1 cm.

the *z*-axis divided by the CTDI free-in-air value measured at the same scanning parameters (dose values for the second and third trimesters of gestation can be found in the article published by Damilakis *et al* (2010)). The ratio $\frac{\mathrm{NCD}}{\mathrm{NCD}_0}$ can be obtained from figure 5.5.

Dose data provided for the Siemens Sensation 16 CT model can be applied to other scanners using the following formula

$$(\mathrm{NCD})_x = (\mathrm{NCD})_S \cdot \frac{\left(\frac{\mathrm{CTDI}_w}{\mathrm{CTDI}_F}\right)_x}{\left(\frac{\mathrm{CTDI}_w}{\mathrm{CTDI}_F}\right)_S}$$

where *S* refers to the Siemens Sensation 16 scanner and *x* to the scanner of interest.

Example

A pregnant patient in her fifth month of gestation was subjected to a pelvic CT examination to rule out appendicitis. The maternal perimeter was 107 cm. The CT examination was obtained with 120 kV, 1.38 pitch and 20 mm beam collimation using a General Electric Lightspeed 16-slice CT scanner. The Automatic Exposure Control was activated during the examination and the average modulated mAs was 200 mAs. The CTDI_F of the Lightspeed scanner measured at 120 kVp, 20 mm beam collimation and 100 mAs was 25.1 mGy, and the CTDI_w of the same scanner at 120 kVp, 20 mm beam collimation and 100 mAs was 9.9 mGy. The CTDI_F of the

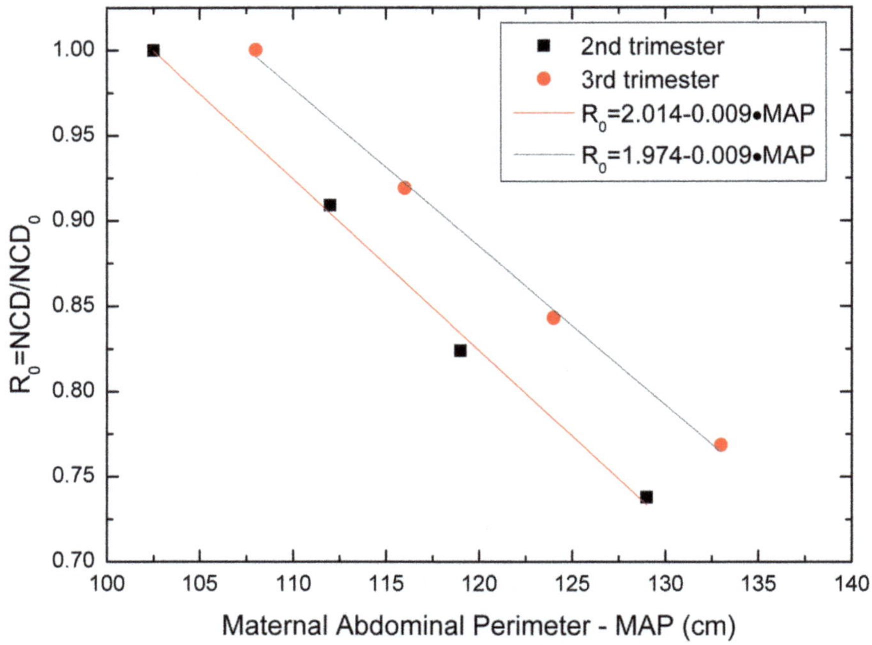

Figure 5.5. The linear relationship between $\frac{NCD}{NCD_0}$ and maternal abdominal perimeter for the phantoms of different sizes representing pregnancy during the second (R_0 = 2.014–0.049 MAP) and third trimesters (R_0 = 1.974–0.009 MAP) of gestation.

Sensation 16 scanner measured at 120 kVp and 100 mAs is 17.1 mGy, and the $CTDI_w$ of the same scanner at 120 kVp and 100 mAs is 7.6 mGy.

Conceptus dose D_c is estimated using the following equation

$$D_c = \left(\frac{CTDI_F\,(mGy)}{100\ mAs} \cdot mAs\right) \cdot NCD(p_0) \cdot \left(\frac{NCD}{NCD_0}\right)_{p_x} \cdot \frac{\frac{CTDI_{w(GE)}}{CTDI_{F(GE)}}}{\frac{CTDI_{w(S)}}{CTDI_{F(S)}}}$$

$NCD(p_0)$ is estimated taking into consideration beam collimation, pitch and data from table III, page 6415, Damilakis *et al* (2010) ($z_1 = -10$ to $z_2 = +8$, 120 kV, BC = 24 mm)

$$NCD(p_0) = \frac{20\ mm}{20\ mm} \cdot \frac{1}{1.38} \cdot 0.4789 = 0.347\ mGy\ mGy^{-1}$$

Data of figure 5.5 are used to calculate $\left(\frac{NCD}{NCD_0}\right)_{p_x}$

$$\left(\frac{NCD}{NCD_0}\right)_{p_x} = 2.014 - 0.009 \cdot 107 = 1.051$$

Therefore,

$$D_c = \frac{25.1\ \text{mGy}}{100\ \text{mAs}} \cdot 200\ \text{mAs} \cdot 0.347 \cdot 1.051 \cdot 0.888 = 16.3\ \text{mGy}.$$

Damilakis *et al* (2010)

Monte Carlo simulations are performed using either voxelized phantoms or patient models. Patient models are based on images of patients who have previously undergone CT examinations and, therefore, are characterized with different body statures representing different human body anatomies and sizes. On the other hand, voxelized phantoms represent only standard size patients of standard ages. Most Monte Carlo approaches have been based on geometrical phantoms or on voxelized phantoms.

A methodology has been developed by Damilakis *et al* (2010) for the estimation of conceptus dose from CT examinations performed on pregnant patients during the first 7 weeks of gestation. In this study, a patient-specific Monte Carlo set of simulation experiments was performed on 117 patient models to develop a technique for conceptus dose estimation from abdominal and pelvic CT examinations. The CT scanner modeled was a Siemens Sensation 16.

The following equation provides the relationship among conceptus dose D (in mGy), body perimeter (P) (in cm) and skin-to-conceptus distance (in cm)

$$D = \text{CTDI}_F \cdot [(1.179 - 0.0043) \cdot (P) - 0.0238 \cdot (\text{SCD})] \cdot \frac{\left(\frac{\text{CTDI}_W}{\text{CTDI}_F}\right)_x}{\left(\frac{\text{CTDI}_W}{\text{CTDI}_F}\right)_\text{S}}$$

The $(\text{CTDI}_W/\text{CTDI}_F)_x$ ratio refers to the scanner where the examination was performed, whereas the $(\text{CTDI}_W/\text{CTDI}_F)_\text{S}$ ratio to the Siemens Sensation 16 CT scanner. The CTDI_F of the CT scanner should be obtained at the same kV and mAs selected for the abdomen and pelvis examination.

Example

A young female patient was admitted to an emergency room complaining of abdominal pain following a car accident. An abdominal CT including the pelvis was requested by the clinicians. The CT examination was obtained with 120 kV, 1.00 pitch and 20 mm beam collimation using a General Electric Lightspeed 16-slice CT scanner. The Automatic Exposure Control was activated during the examination and the average modulated mAs was 350 mAs. After the examination, the patient was diagnosed as pregnant at her 6th week of gestation. By means of an ultrasound unit, the conceptus depth was found to be 14 cm. The maternal perimeter was 85 cm. The CTDI_F of the Lightspeed scanner measured at 120 kVp, 20 mm beam collimation and 100 mAs was 25.5 mGy, and the CTDI_W of the same scanner at 120 kVp, 20 mm beam collimation and 100 mAs was 9.9 mGy. The $\text{CTDI}_W/\text{CTDI}_F$ ratio for the Siemens Sensation 16 scanner is 0.444.

The $CTDI_W/CTDI_F$ ratio for the General Electric Lightspeed 16 scanner is 9.9/25.5 = 0.388 and, therefore, the ratio of the $CTDI_W/CTDI_F$ ratios of the two scanners is 0.388/0.444 = 0.874.

Using the above data, the dose to the conceptus is given by

$$D = \frac{25.5\,\text{mGy}}{100\,\text{mAs}} \cdot 350\,\text{mAs} \cdot (1.179 - 0.0043 \cdot 85 - 0.0238 \cdot 14) \cdot 0.874 = 37.5\,\text{mGy}$$

A disadvantage of the above conceptus dose calculation is that it cannot be used when (a) the conceptus is outside the field of view of the CT examination and (b) the extent of scanning volume differs considerably from that of a typical abdominopelvic examination. Moreover, this technique estimates conceptus dose from examinations performed during early gestation.

Angel *et al* (2008)

A method for the estimation of conceptus dose from CT examinations has been published in *Radiology* by Angel *et al.* Twenty-four patient models selected to represent a range of gestational ages were created using image data from pregnant patients who had undergone abdominopelvic CT examinations. The authors found the following relationship between fetal dose and the maternal perimeter:

$$\text{ND}_c = -0.122 \cdot P + 23.11$$

where ND_c is the normalized conceptus dose (mGy 100 mAs^{-1}) and P the maternal perimeter. The authors also provided a two-variable equation which correlates conceptus dose in mGy 100 mAs^{-1} with conceptus depth DE_c and the perimeter of the mother P:

$$\text{ND}_c = -0.119 \cdot P - 0.29 \cdot \text{DE}_c + 24.56$$

This method does not take into account variations in scan length and is not applicable if the conceptus is partly included in the imaging volume or if it is remote from the directly exposed tissues. Moreover, only one CT scanner model, GE Lightspeed 16, was used.

Example

An abdominopelvic CT scan was prescribed for a pregnant patient with a perimeter of 107 cm. The CT examination was obtained with 120 kV, 200 mAs, 1.38 pitch using a General Electric Lightspeed 16-slice CT scanner. The normalized conceptus dose (mGy 100 mAs^{-1}) is given by

$$\text{ND}_c = \frac{23.11 - 0.122 \cdot 107}{1.38} = 7.29 \frac{\text{mGy}}{100\,\text{mAs}}$$

The dose to the fetus is

$$\text{D}_c = 14.6\,\text{mGy}$$

5.1.3 Conceptus dose estimation (CoDE) software

A software package CoDE has been developed to (a) calculate conceptus doses and risks associated with imaging examinations performed on the expectant mother

Table 5.1. Radiographic examinations.

Body region under examination	Projections[a]	Field size (cm^2)	Tube voltage (kVp)	Total filtration (mm Al)
Abdomen	AP	36 × 43	50–120	2.5–5.0
	PA	36 × 43	50–120	2.5–5.0
Chest	AP	36 × 43	50–120	2.5–5.0
	PA	36 × 43	50–120	2.5–5.0
	LAT	36 × 43	50–120	2.5–5.0
Kidneys	AP	36 × 24	50–120	2.5–5.0
	PA	36 × 24	50–120	2.5–5.0
Lumbar spine	AP	36 × 43	50–120	2.5–5.0
	LAT	36 × 43	50–120	2.5–5.0
	LAT LSJ	18 × 24	50–120	2.5–5.0
	LPO	24 × 30	50–120	2.5–5.0
	RPO	24 × 30	50–120	2.5–5.0
Pelvis/colon	AP	36 × 43	50–120	2.5–5.0
	PA	36 × 43	50–120	2.5–5.0
	AP hip joint	24 × 35	50–120	2.5–5.0
Thoracic spine	AP	24 × 43	50–120	2.5–5.0
	LAT	20 × 49	50–120	2.5–5.0
Urinary bladder	AP	24 × 21	50–120	2.5–5.0

[a] Angulations of all projections refer to the position of the X-ray tube with respect to the patient's body.

and (b) anticipate conceptus dose for the pregnant employee who participates in fluoroscopically guided interventional procedures. CoDE has been uploaded to http://embryodose.med.uoc.gr/code/ and its use is free of charge.

The CoDE tool user interface comprises of four modules: (1) a radiography, (2) a fluoroscopy and (3) a computed tomography conceptus dose/risk calculation module for radiological examinations performed on pregnant patients as well as (4) an occupational dose estimation module for conceptus exposure calculations.

The theoretical radiogenic risk for childhood cancer associated with in-utero exposure is based on a risk coefficient of 1.2×10^{-2} per Gy as recommended by the International Commission on Radiological Protection (ICRP 2003).

Conceptus dose estimation for examinations performed on pregnant patients using CoDE is possible for the first, second and third trimester of gestation for various radiographic projections, kVp and total filtration values (table 5.1).

The CoDE fluoroscopy module provides estimates for the conceptus absorbed dose and associated risk for childhood cancer from gastrointestinal (table 5.2), cardiac (table 5.3), orthopedic (table 5.4) and other (table 5.5) fluoroscopically guided procedures performed on the pregnant patient. Dose calculations are possible for all trimesters of gestation for various kVp and total filtration values.

The CoDE CT module provides estimates for the conceptus absorbed dose and associated risk for childhood cancer from computed tomography examinations of

Table 5.2. Gastrointestinal fluoroscopically guided procedures.

Procedure	Projection[a]	Field size (cm^2)	Tube voltage (kVp)	Total filtration (mm Al)
Barium enema	AP colon/pelvis	24 × 30	50–120	2.5–9.0
	AP rectum	15 × 15	50–120	2.5–9.0
	LAO colon	35 × 35	50–120	2.5–9.0
	LAO flexure	28 × 35	50–120	2.5–9.0
	LAT rectum	17 × 17	50–120	2.5–9.0
	LPO colon	35 × 35	50–120	2.5–9.0
	LPO rectum	19 × 19	50–120	2.5–9.0
	PA pelvis/colon	40 × 40	50–120	2.5–9.0
	PA rectum	15 × 15	50–120	2.5–9.0
	RAO colon	35 × 35	50–120	2.5–9.0
	RAO flexure	28 × 35	50–120	2.5–9.0
	RAO rectum	19 × 19	50–120	2.5–9.0
Barium follow through	AP small intestine	22 × 18	50–120	2.5–9.0
	PA small intestine	22 × 18	50–120	2.5–9.0
Barium meal	AP duodenum	15 × 15	50–120	2.5–9.0
	AP stomach	18 × 22	50–120	2.5–9.0
	AP upper stomach	15 × 15	50–120	2.5–9.0
	LAO stomach	21 × 21	50–120	2.5–9.0
	LAT stomach	19 × 24	50–120	2.5–9.0
	LPO duodenum	19 × 19	50–120	2.5–9.0
	LPO stomach	21 × 21	50–120	2.5–9.0
	PA duodenum	15 × 15	50–120	2.5–9.0
	PA stomach	18 × 21	50–120	2.5–9.0
	PA upper stomach	15 × 15	50–120	2.5–9.0
	RAO duodenum	19 × 19	50–120	2.5–9.0
	RAO stomach	21 × 21	50–120	2.5–9.0
Barium swallow	LAO esophagus	13 × 47	50–120	2.5–9.0
	LAT throat	18 × 24	50–120	2.5–9.0
	LPO esophagus	13 × 47	50–120	2.5–9.0
	RAO esophagus	13 × 47	50–120	2.5–9.0

[a] Angulations of all projections refer to the position of the tube with respect to the patient's body.

the trunk performed on the pregnant patient. A pull-down menu guides the user to select the gestational stage of the exposed pregnant individual that is 0–7 weeks, 8–12 weeks, 13–25 weeks and 26–40 weeks. The user has to define the boundaries of the scanned volume (the start and the end of the scan) moving the cursors appropriately at the corresponding anatomical locations on the provided trunk scout view.

CoDE also provides prospective or retrospective estimates of absorbed dose to the embryo of a pregnant employee involved in occupational exposure from fluoroscopically guided interventional procedures. Table 5.6 shows the 17 fluoroscopic projections for which dose estimation is possible by using CoDE.

Table 5.3. Cardiac fluoroscopically guided procedures.

Procedures	Projection[a]	Field size (cm^2)	Tube voltage (kVp)	Total filtration (mm Al)
Coronary angiography/	PA	12.5 × 12.5	70–100	3–13
Angioplasty	PA CRANIAL 30	12.5 × 12.5	70–100	3–13
	PA CAUDAL 30	12.5 × 12.5	70–100	3–13
	LLAT	12.5 × 12.5	70–100	3–13
	RAO 30	12.5 × 12.5	70–100	3–13
	LAO 40	12.5 × 12.5	70–100	3–13
	LAO 45 CRANIAL 20	12.5 × 12.5	70–100	3–13
	RAO 20 CRANIAL 20	12.5 × 12.5	70–100	3–13
	RAO 20 CAUDAL 20	12.5 × 12.5	70–100	3–13
	LAO 40 CAUDAL 30	12.5 × 12.5	70–100	3–13
Pacemaker implantation	PA	12.5 × 12.5	50–120	2.5–9.0
	RAO 30	14 × 14	50–120	2.5–9.0
	LAO 30	14 × 14	50–120	2.5–9.0
Cardiac ablation	AP	12.5 × 12.5	50–120	2.5–9.0
	LAO 30	14 × 14	50–120	2.5–9.0
	RAO 30	14 × 14	50–120	2.5–9.0
	Guidance iliac	6 × 6	70–100	3–13
	Guidance jugular	6 × 6	70–100	3–13

[a] Angulations of all projections refer to the position of the image intensifier with respect to the patient's body.

Table 5.4. Orthopedic fluoroscopically guided procedures.

Procedures	Projection[a]	Field size (cm^2)	Tube voltage (kVp)	Total filtration (mm Al)
Femoral fractures	Hip joint LAT	15 × 15	70–100	3–13
	Hip joint PA	15 × 15	70–100	3–13
Kyphoplasty	AP lumbar spine	8 × 15	70–100	3–13
	LAT lumbar spine	8 × 15	70–100	3–13

[a] Angulations of all projections refer to the position of the image intensifier with respect to the patient's body.

When all the necessary data has been supplied, the dose absorbed by the conceptus of the pregnant employee is calculated and presented in the corresponding field. The scatter exposure map corresponding to the selected fluoroscopic projection is also presented (figure 5.6).

5.2 Methods to calculate paediatric dose from diagnostic and interventional procedures

5.2.1 Introduction

Radiologists constantly face the dilemma of trying to lower patient dose whenever this is possible, while at the same time keeping the image quality adequate to provide

Table 5.5. Other fluoroscopically guided procedures.

Procedures	Projection[a]	Field size (cm^2)	Tube voltage (kVp)	Total filtration (mm Al)
Endoscopic retrograde cholangio-pangratography (ERCP)	LLAT abdomen	20 × 20	80–100	2.5–9.0
Inferior vena cava filter	Guidance iliac	6 × 6	70–100	3–13
placement	Guidance jugular	6 × 6	70–100	3–13
	Suprenal placement	15 × 8	80–100	3–13
	Subrenal placement	15 × 8	80–100	3–13
Cysteourethrography	AP[T] bladder	24 × 21	50–120	2.5–9.0
Prophylactic	PA[T] left artery	18 × 22.5	80–100	3–13
hypogastric artery balloon	PA[T] right artery	18 × 22.5	80–100	3–13
occlusion (HABO)	RA 20 left artery	18 × 22.5	80–100	3–13
	RAO 20 right artery	18 × 22.5	80–100	3–13
	LAO 20 left artery	18 × 22.5	80–100	3–13
	LAO 20 right artery	18 × 22.5	80–100	3–13

[a] Angulations of all projections refer to the position of the image intensifier with respect to the patient's body.
[T] The angulation of these projections refers to the position of the tube with respect to the patient's body.

an accurate diagnosis. The measurement or estimation of patient doses is thus very important for control and optimization and even a 10% reduction in dose is a worthwhile objective of dose optimization.

Patient dosimetry in paediatric patients undergoing diagnostic radiology examinations require special attention due to a number of reasons such as (1) their higher risk from radiation, (2) longer life expectancy, (3) special measuring equipment that may be needed, (4) limited number of facilities performing these examinations, (5) possible lack of trained staff to perform the examinations, etc. Previous chapters have described in detail the various equipment or software that can be used to measure or calculate paediatric radiation dose in the everyday clinical routine. The following paragraph will attempt to bring all of the information together with the objective to calculate paediatric radiation dose in general radiography, fluoroscopy and CT.

5.2.2 Methodology to calculate paediatric dose in radiography and fluoroscopy

Dosimetry measurements using phantoms are useful for quality control and comparison between X-ray machines, X-ray protocols, hospitals, etc. They cannot provide a direct estimate of the radiation dose for a particular patient. Patient measurements are required in order to analyze dose variations seen in practice due to differences in patient size, technique or skill between operators, establish DRLs, etc. Table 5.7 provides the main dose quantities applied for dose measurements in radiography, fluoroscopy and CT both for phantom and patient dosimetry (adapted from table 8.1, page 106, IAEA report 457 (IAEA 2007)). The paragraphs below

Table 5.6. The fluoroscopic projections investigated.

Projection	Abbreviation	
Posterior-anterior	PA	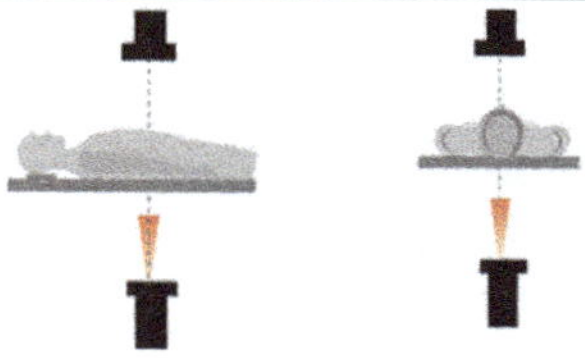
Posterior-anterior/caudal 30°	PA/CAU 30	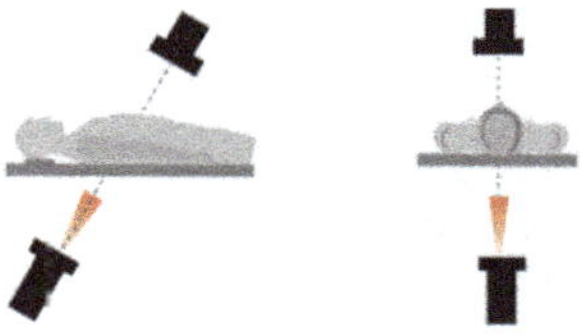
Posterior-anterior/cranial 30°	PA/CRA 30	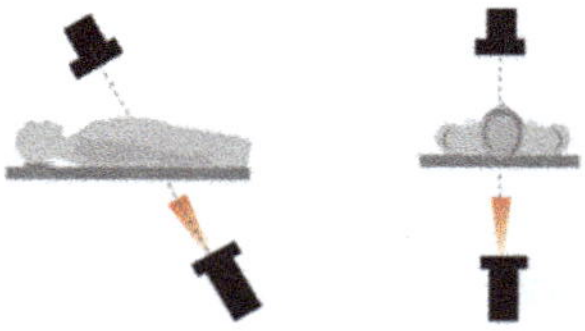
Right anterior oblique 30°	RAO 30	
Left anterior oblique 30°	LAO 30	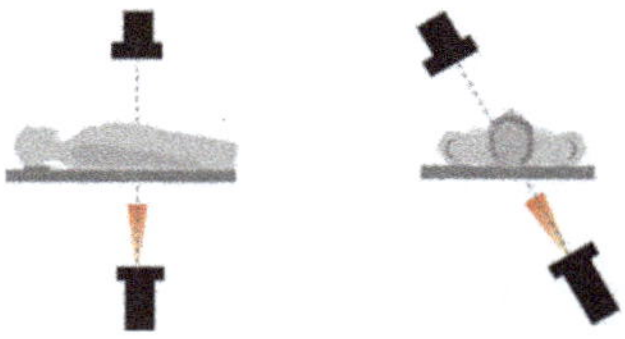
Left anterior oblique 45°	LAO 45	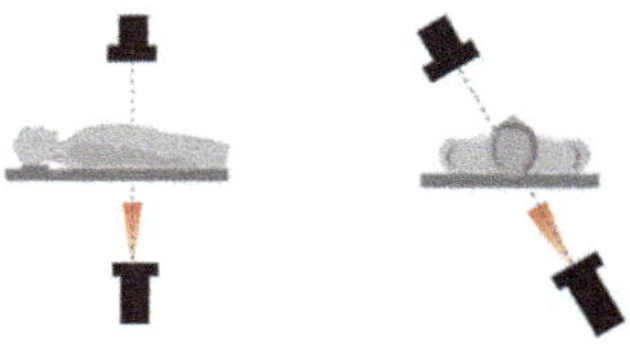
Right anterior oblique 45°	RAO 45	
Left lateral	LLAT	
Right lateral	RLAT	

Left anterior oblique 40°/caudal 25°	LAO 40/CAU 25	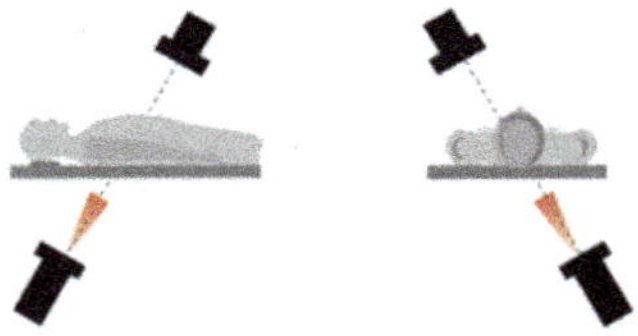
Right anterior oblique 40°/caudal 25°	RAO 40/CAU 25	
Left anterior oblique 40°/cranial 25°	LAO 40/CRA 25	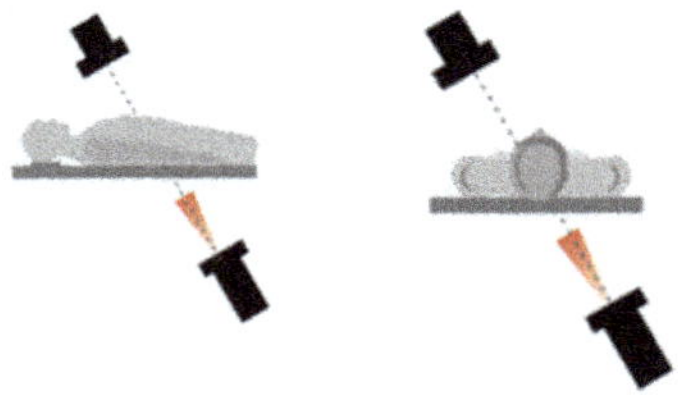
Right anterior oblique 40°/cranial 25°	RAO 40/CRA 25	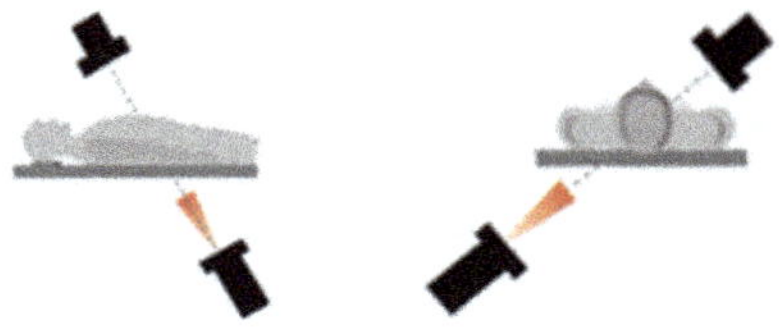
Left anterior oblique 20°/caudal 20°	LAO 20/CAU 20	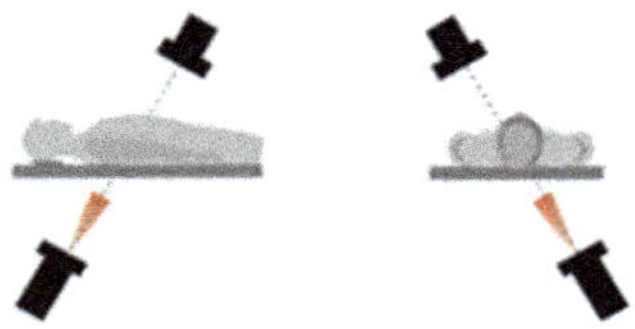
Right anterior oblique 20°/caudal 20°	RAO 20/CAU 20	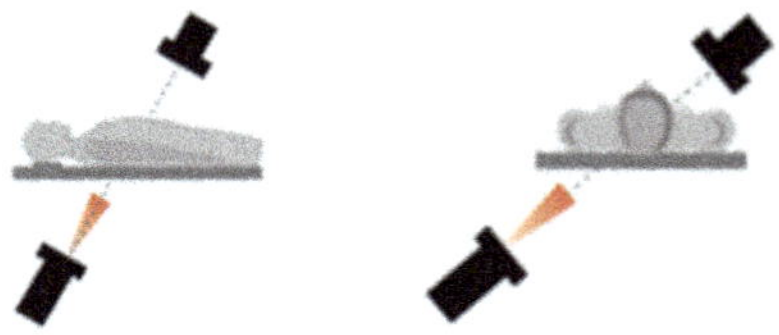
Left anterior oblique 20°/cranial 20°	LAO 20/CRA 20	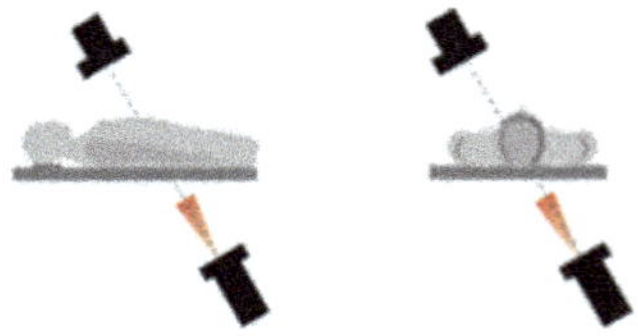
Right anterior oblique 20°/cranial 20°	RAO 20/CRA 20	

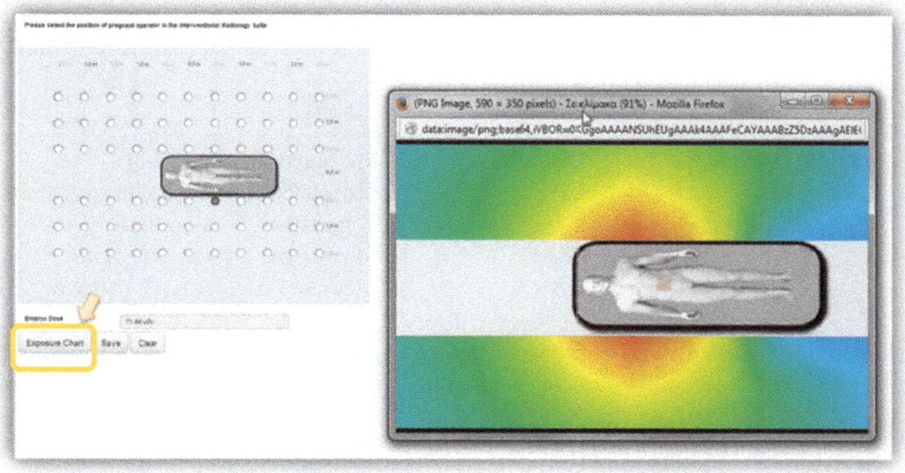

Figure 5.6. Scatter exposure map obtained for a fluoroscopic projection using CoDE. Exposure chart (arrow) can be saved for future use.

describe the methods to estimate or measure patient dose in clinical routine. The reader can also visit the respective sections in IAEA report 457 for more details (IAEA 2007).

Measurement of KAP

In the case the X-ray machine is equipped with a KAP meter, the value of the KAP can be recorded (the KAP meter should have a digital resolution of 0.1 $mGym^2$ or better to meet the need for low paediatric exposures). The values can be also stored in the Digital Imaging and Communications in Medicine (DICOM) structures for further use or archive. KAP value displays should be verified with an external calibrated KAP meter. Detailed methodology for KAP calibration can be found either in the NRPB National protocol for patient dose measurements in diagnostic radiology (NRPB 1992), or in the IAEA report 457 (IAEA 2007). Calibration should ideally be done annually. This is the most direct, easy way to measure patient dose in the everyday clinical routine in radiography and fluoroscopy.

Measurement of ESAK

ESAK provides direct assessment of patient dose and can be measured by thermoluminescence dosimeters or optically stimulated luminescence dosimeters (OSLDs). Details on the use of these dosimeters can be found in section 1.2.

Measurement of IAK and X-ray tube output

IAK is estimated indirectly from the X-ray tube output at the selected distance and exposure parameters using the inverse square law following the specific steps below:

1. Select type of examination and weight range to be investigated.
2. Position the patient and X-ray equipment for the desired examination and set the appropriate exposure parameters.

Table 5.7. Main dose quantities applied for dose measurements in radiography, fluoroscopy and CT both for phantom and patient dosimetry (adapted from table 8.1, page 106, IAEA report 457 (IAEA 2007)).

Modality	Measurement/ subject	Measured quantity	Comments
Radiography	Phantom	IAK	Measured on a phantom
	Patient	IAK	Calculated from exposure factors and measured X-ray output
		ESAK	Measurement on patient skin
		KAP	Measured using a KAP meter or taken from the X-ray console
Fluoroscopy	Phantom	Entrance surface air kerma rate	Measured directly on a phantom or calculated from the incident air kerma rate using backscatter factors
	Patient	KAP	Measured using a KAP meter or taken from the X-ray console
CT	Phantom	CTDI quantities	Measurements in air or in PMMA head and body phantoms
	Patient	CTDI quantities	Taken from the CT console
		DLP	Taken from the CT console or calculated from scan length and CTDI per rotation

3. Measure patient thickness at the center of the beam.
4. Expose the patient and record the exposure parameters.
5. IAK can be determined by calculation from recorded exposure parameters and measured tube output. For selected filtration, the X-ray tube output can be measured for a representative set of tube voltages and tube loadings that adequately sample the patient exposure parameters used in radiography. A detailed description of output measurements can be found also in IAEA report 457 (IAEA 2007). ESAK can be obtained from IAK using tabulated backscatter factors.

Specifically for angiography X-ray machines, the console occasionally displays the Interventional Reference Point Kerma or Cumulative Air Kerma as named by some manufacturers. The particular value can be used in order to provide an estimate of ESAK. Both the Interventional Reference Point Kerma and Cumulative Air Kerma can provide an estimate of the cumulative skin dose.

5.2.3 Methodology to calculate paediatric dose in CT

$CTDI_{vol}$ is measured with a calibrated pencil ionization chamber specially designed for CT and an electrometer. For free-in-air measurements only the pencil chamber is needed. For CT measurements in PMMA, a standard head 16 cm CT dosimetry phantom is employed to simulate the paediatric patient. One must keep in mind though that the use of this phantom will lead to underestimation for very small children and infants that have body size much smaller than 16 cm diameter. The measurements must be performed for the paediatric examination protocols in use. DLP is easily calculated from $CTDI_{vol}$ by multiplying $CTDI_{vol}$ with the length of CT exam. SSDE can be further calculated with the use of look-up tables based on patient thickness from the AAPM reports 204 (AAPM 2011) and 220 (AAPM 2014).

5.2.4 Paediatric organ dose and effective dose; practical issues

IAK, ESAK or KAP of radiographic or fluoroscopic procedures can be multiplied with specific conversion factors (CFs) that can be found in the literature to calculate equivalent organ dose and *E*. IAEA report 24 (2014) summarizes publications that could be used to derive CF for radiography. CFs are also defined using Monte Carlo algorithms (there are many types of commercial or in-house software for this purpose such as PCXMC, CT Expo, ImPACT CT, NRPB data sets (Jones and Schrimpton, 1991), etc. These are described in detail in section 1.4) or using phantoms and special dosimeters. The literature has plenty of publications that provide CFs for a specific patient size and X-ray examination. Regarding skin doses that are occasionally of concern in complex fluoroscopic procedures, smaller paediatric patients receive much lower radiation dose rate to the entrance of the skin. Consequently, cumulative skin dose to the paediatric patient is lower compared to an adult for fluoroscopy or interventional procedures. This practically eliminates the deterministic effects for children in fluoroscopic examinations. Thus, skin dose radiation effects should not be considered for young children, provided of course that there is not a technical issue or malfunction with the X-ray equipment and that appropriate paediatric protocols are applied at all times. This fact may change at an older age when the size of the child may reach the size of an adult. Evidently, the risk from radiation should be focused on stochastic effects due to their higher radio sensitivity and their longer expected lifetime.

As far as *E* is concerned, one should always remember that *E* is designed for protection purposes for whole body irradiation of populations (both males and females). It should not be used to estimate the risk for incidence of cancer and/or heritable effects for a specific patient. As a consequence, it should only be used to compare difference radiological examinations in terms of radiation dose or compare practices of a particular exam in different hospitals or health centers. If one finally decides to estimate *E* (for reasons mentioned above) and does so using commercial software described in section 1.4 then the tissue weighting factors used by the program should also be checked.

5.2.5 Examples in everyday clinical practice

The reader can find below a few examples on paediatric dose estimation using a reference from the literature. The solution offered is not the only one; it just provides an example of how one can estimate patient dose in everyday clinical practice.

Radiography

Question: A 3-year-old girl was brought to a hospital with nontrivial head trauma. Skull PA and LAT radiographs were performed applying 70 kV using an X-ray radiography machine with a 4 mm beam filtration. Entrance surface dose was 0.45 mGy for the PA skull and 0.55 mGy for the lateral radiograph. Which is the total effective dose to the patient?

Reference used: Stratakis *et al* 2005.

Calculations: Table 6 from the publication above is needed to find CFs for the particular kV and HVL.

Effective dose in PA = 16 μSv mGy^{-1} * (0.45 mGy) = 7.2 μSv

Effective dose in LAT = 20 μSv mGy^{-1} * (0.55 mGy) = 11.0 μSv

E = (7.2 + 11.0) μSv = 18.2 μSv.

CT

Question: A 5-year-old boy without personal or familial history has arrived to the CT department with an acute headache and loss of consciousness during a basketball competition. Upon waking up, the child had right hemiplegia, aphasia and left Horner syndrome. A full scan head CT scan at 120 kV and 228 mAs was performed with a CTDI and a DLP of 26 mGy and 348 mGycm, respectively. Estimate the effective dose.

Reference used: Bongartz *et al* 2004.

Calculations: Table 4 from the aforementioned publication provides normalized values of effective dose per DLP for the particular page in question.

E = 0.0040 * (348 mSv $mGycm^{-1}$) = 1.39 mSv.

Fluoroscopy

Question: A 1-year-old girl with abdominal pain has undergone a barium meal examination with a KAP of 2 $mGycm^2$. Can you estimate the stomach and the breast dose as well as the effective dose to the patient?

Reference used: Dimitriadis *et al* 2011.

Calculations: The particular publication has various CFs. Table 3 provides the CVs for stomach and breast equivalent dose and also the CV for effective dose estimation. The multiplication of CV to KAP value provides equivalent dose and E:

	CV	KAP	Result (μSv)
Stomach dose	5.24	2	10.48
Breast dose	1.57	2	3.14
Effective dose	1.67	2	3.34

Interventional cardiology

Question: A 5-year-old girl underwent atrial septal defect (ASD) treatment with a KAP product of 6 $Gycm^2$. Can you estimate the effective dose?

Reference used: Karambatsakidou *et al* 2009.

Calculations: ASD is a therapeutic procedure. Figure 3 of the publication above can provide the mean CVs required for our estimations; for the 5-year-old patient and for a therapeutic procedure such as ASD, the CV is 0.9 mSv $Gycm^{-2}$. Thus:

Effective dose = 0.9 * 6 mSv = 5.4 mSv.

References

Angel E, Wellnitz C and Goodsitt M *et al* 2008 Radiation dose to the fetus for pregnant patients undergoing multidetector CT imaging: Monte Carlo simulations estimating fetal dose for a range of gestational age and patient size *Radiology* **249** 220–7

AAPM Task Group 204 Report 2011 Size-Specific Dose Estimates (SSDE) in Pediatric and Adult Body CT Examinations

AAPM Task Group Report 220 2014 Use of Water Equivalent Diameter for Calculating Patient Size and Size-Specific Dose Estimates (SSDE) in CT

Bongartz G *et al* 2004 European guidelines for multislice computed tomography. Funded by the European Commission. Contract number FIGM-CT2000–20078-CT-TIP (http://msct.eu/CT_Quality_Criteria.htm)

Damilakis J, Theoharopoulos N, Perisinakis K, Manios E, Dimitriou P, Vardas P and Gourtsoyiannis N 2001 Embryo radiation dose and risk from cardiac catheter ablation procedures *Circulation* **104** 893–7

Damilakis J, Tzedakis A, Sideri L, Perisinakis K, Stamatelatos I and Gourtsoyiannis N 2002 Normalized conceptus doses for abdominal radiographic examinations calculated using a Monte Carlo technique *Med. Phys.* **29** 2641–8

Damilakis J, Perisinakis K, Tzedakis A, Papadakis A and Karantanas A 2010 Radiation dose to the embryo from NDCT during early gestation: A method that allows for variations in maternal body size and embryo position *Radiology* **257** 483–9

Damilakis J, Tzedakis A, Perisinakis K and Papadakis A E 2010 A method of estimating embryo doses resulting from multidetector CT examinations during all stages of gestation *Med. Phys.* **37** 6411–20

Dimitriadis A, Gialousis G, Makri T, Karlatira M, Karaiskos P, Georgiou E, Papaodysseas S and Yakoumakis E 2011 Monte Carlo estimation of radiation doses during paediatric barium meal and cystourethrography examinations *Phys. Med. Biol.* **56** 367–82

IAEA Technical Reports Series No. 457 2007 *Dosimetry in Diagnostic Radiology: An International Code of Practice* (Vienna, MD: IAEA)

ICRP 2003 Biological effects after prenatal irradiation (embryo and fetus) ICRP Publication 90 *Ann. ICRP* **33** 1–2

ICRU Report 74 2006 *Patient Dosimetry for X Rays Used in Medical Imaging* (Bethesda, MA: ICRU)

IAEA 2014 Dosimetry in diagnostic radiology for paediatric patients *Human Health Series* 24 (Vienna: IAEA)

Jones D G and Shrimpton P C 1991 Survey of CT practice in the UK. Part 3: Normalised organ doses calculated using Monte Carlo techniques. NRPB-R250, Chilton

Karambatsakidou A, Sahlgren B, Hansson B, Lidegran M and Fransson A 2009 Effective dose conversion factors in paediatric interventional cardiology *Br. J. Radiol.* **82** 748–55

NRPB 1992 *Dosimetry Working Party of the Institute of Physical Sciences in Medicine. National Protocol for Patient Dose Measurements in Diagnostic Radiology* (Chilton, MD: NRPB)

Samara E, Stratakis J, Enele Melono J M, Mouzas I A, Perisinakis K and Damilakis J 2009 Therapeutic ERCP and pregnancy: is the radiation risk for the embryo trivial? *Gastrointest. Endosc.* **69** 824–31

Solomou G, Perisinakis K, Tsetis D, Stratakis J and Damilakis J 2016 Data and methods to estimate fetal dose from fluoroscopically guided prophylactic hypogastric artery balloon occlusion *Med. Phys.* **43** 2990–7

Stratakis J, Damilakis J and Gourtsoyiannis N 2005 Organ and effective dose conversion coefficients for radiographic examinations of the pediatric skull estimated by Monte Carlo methods *Eur. Radiol.* **15** 1948–58

IOP Publishing

Radiation Dose Management of Pregnant Patients, Pregnant Staff and Paediatric Patients
Diagnostic and interventional radiology
John Damilakis

Chapter 6

Optimization of radiological examinations performed during pregnancy

Kostas Perisinakis and John Stratakis

Optimization of radiological examinations on pregnant patients aims at minimization of the radiation burden to both the expectant mother and the conceptus without impairing the quality of the diagnostic information expected from the examination. Given that radiological examinations performed on pregnant patients with the use of modern radiological equipment may be of high complexity, optimization may be a difficult task that requires familiarization with the specific technological features of the equipment used and close collaboration of the radiologist with the medical physicist and the operator. Some general guidelines and practical hints for the optimization of radiologic procedures performed on pregnant patients are discussed in this chapter.

6.1 Radiography/fluoroscopy during pregnancy: methods for dose optimization

6.1.1 Introduction

According to the International Commission on Radiological Protection, 'prenatal doses from most correctly performed diagnostic procedures present no significantly increased risk of prenatal or postnatal death, developmental damage including malformation, or impairment of mental development over the background incidence of these entities; life-time cancer risk following in utero exposure is assumed to be similar to that following irradiation in early childhood' (ICRP 2007). However, the As Low As Reasonably Achievable (ALARA) principle denotes that the use of radiation may be substituted with diagnostic methods without ionizing radiation whenever they are equivalent in reaching the same diagnosis, as well that every medical diagnostic imaging examination must be justified; this means that its diagnostic benefits for a specific patient in a specific clinical situation must exceed its risks. Justification is based on the specific benefits and risks for both mother and child. The stronger the arguments for a critical situation of one of them are, the easier the justification; in contrast, a vague suspicion would not justify an important exposure.

Radiation-related risks are present throughout gestation. Knowledge of risks to the conceptus has a wide range of uncertainty. Effects may be either stochastic (cellular mutation, with no threshold and a probability of damage proportional to the dose; e.g. cancer induction) or deterministic (multicellular injury, e.g. malformations), with a rather high threshold. The magnitude of these risks is highly dependent on the gestational age during which exposure takes place and the conceptus absorbed dose. Biologic systems with a high fraction of proliferating cells demonstrate high radiation receptiveness. Radiation risks are most significant during preimplantation and organogenesis and portions of the first trimester, slightly less in the second trimester and least in the third trimester. There is no evidence that radiation dose in the diagnostic ranges is associated with an increased occurrence of congenital malformation, miscarriage, or mental disability.

6.1.2 Radiation protection approach for imaging the female patient

Every woman of child-bearing age (typically ages 13–55 years) has to be considered as potentially pregnant and should be asked whether she is pregnant or thinks she could be. The pregnancy status becomes critically important before imaging the abdomen and pelvis. While queries about the potential for pregnancy and the menstrual cycle are important, whenever the answer is not clearly negative or even positive, a pregnancy test should be done. Unless for lifesaving procedures, the pregnancy status must be established before the uterus is exposed to direct radiation when the conceptus dose is likely to exceed 1 mGy.

Once pregnancy has been confirmed, justification will differ whether the conceptus will receive direct ionizing radiation. Imaging of the trunk of the body without direct radiation to the conceptus (e.g. chest radiography during the first two trimesters) will cause some scatter radiation and a dose below 1 mGy to the conceptus. Examinations of this category will usually be performed when they are well-justified, but the following step, which is optimization of the procedure, becomes more important.

In the situation of imaging the abdomen or pelvis with direct radiation to the conceptus, alternatives using ultrasound or MRI or avoiding imaging should be sought. When they are unavailable or have been ineffective, an examination using ionizing radiation will be considered depending on the individual medical benefit and the degree of urgency. In addition, the procedure can be postponed until after the critical phase of organogenesis, or even after the pregnancy. Every pregnancy situation deserves individual evaluation, and the decision shall be taken based on the best scientific knowledge.

For an imaging examination using ionizing radiation, obtaining consent from a patient known to be pregnant is an essential component of providing comprehensive medical care. This process requires: (1) a realistic overview of the risk to the patient and her developing child from the examination, and (2) the beneficial role of this imaging procedure in maternal or fetal health evaluation. Whether in written or in verbal consent, this interaction must be documented in the patient's medical record and in compliance with national regulations.

6.1.3 Practices for optimizing conceptus dose in radiographic examinations performed on pregnant patients

For diagnostic radiologic procedures outside the abdomen and pelvis, including head and neck, the chest, and the extremities, the only radiation to which the conceptus is subjected is scattered radiation, which almost always results in a very low dose. When standard precautions are taken to avoid direct irradiation of the abdomen/pelvis using patient positioning and X-ray beam collimation, the dose delivered will not present significant risk to the conceptus. For radiological examinations, the highest radiation exposure to the conceptus occurs when the abdominal/pelvic region is exposed to the primary X-ray beam. Radiation exposure parameters may be reduced and a certain degree of compromise in image quality is tolerable; even so, the quality cannot decrease beyond a certain level required for diagnosis. General protection measures can be utilized to all radiographic examinations when performing such a radiographic examination on a pregnant patient with the radiation field partially or fully irradiating the conceptus region. Adequate filtration (minimum 2.5 mm Al, in general radiology) significantly reduces the patient dose due to low energy X-rays that do not contribute to the image formation. Inappropriate selection of automatic exposure control settings might lead to images that are too light or too dark. Automatic exposure control devices should always be evaluated, especially when the sensitivity of the screen-film combination has been changed. In addition, the correct operation of the automatic exposure control device requires, for each projection, the selection of the chamber or detector closest to the area of interest, so that this area will have the appropriate density. The use of the anti-scatter grid improves the image quality but increases patient dose. It is advisable to evaluate whether the grid is necessary in equipment where its use is optional according to procedure or patient characteristics. Finally, a technique chart specifically for pregnant patient procedures should be placed beside each X-ray control panel for the various projections. For manual exposures, the techniques (kVp, mA and time) should be specified as a function of body part thickness.

6.1.4 Practices for optimizing conceptus dose in fluoroscopic examinations performed on pregnant patients

Procedure related optimization considerations

Optimizing patient or conceptus dose is not the same as minimizing patient or conceptus dose. Some interventional procedures require high-quality images, long exposure time, or both. It is critically important to always attempt to accomplish the maximum possible dose reduction coherent with acceptable image quality. Simple techniques exist that can accomplish this. These include excluding the conceptus from the primary beam path, using reduced dose modes or collimation, as well as proper selection of the numerous technical factors that affect dose. Optimization is possible through appropriate use of the basic features of interventional fluoroscopic equipment and intelligent use of dose-reducing technology. Many technical parameters can be adjusted during the procedure to reduce radiation use or to improve image quality, depending on the demands of the situation.

Interventions in anatomic regions remote from the conceptus, such as the thorax, head, or the extremities, can usually be performed at any time during pregnancy with proper collimation. If the abdomen and/or pelvis is likely to be in the direct beam or proximal to the primary beam, absorbed conceptus doses can come close or exceed many tenths of mGy. In such cases, care should be taken to minimize the absorbed dose to the conceptus and to also acknowledge that the larger body habitus caused by pregnancy will likely cause an increase in skin entrance dose for the mother. The intervention should be designed to reduce overall dose; however, any alterations in the procedure's method should not unjustifiably reduce the diagnostic or interventional value of the practice.

The most important practical actions to control and ultimately constrain the dose to the conceptus when performing fluoroscopically guided interventional procedures are summarized in table 6.1.

During the first trimester, the conceptus dose is dependent on the distance between the conceptus and the maternal skin surface as opposed to the beam entrance. In addition, it has been shown that the conceptus depth is strongly influenced by the fullness of the mother's bladder. Therefore, during the first trimester, the optimal status of the bladder (pre- or post-void) should be determined. For example, it has been shown that if it becomes necessary for a pregnant woman

Table 6.1. Practical actions to limit conceptus doses in IR procedures.

- Exclude the conceptus from the direct beam if possible.
- Keep beam-on time to an absolute minimum.
- Consider use of intravascular ultrasound in place of X-ray for portions of the procedure.
- Allow for posterior-anterior beam projections whenever possible.
- Use low dose rate pulsed fluoroscopy.
- Use last image hold instead of spot fluorographic images for studying evidence and to plan the technique.
- Remember that dose rates will be greater, and dose will accumulate faster in larger patients (such as mid- to late-term pregnant patients).
- Keep the tube current as low as possible by keeping the tube potential elevated to achieve the appropriate compromise between image quality and low patient and conceptus dose.
- Keep the X-ray tube at maximal distance from the patient.
- Keep the image intensifier or flat-panel detector as close to the patient as possible.
- Do not overuse geometric magnification.
- Minimize exposure from digital subtraction angiography (DSA) by using a low frame rate or by limiting the number of DSA runs. It may be possible to substitute fluoroscopic loops for DSA when the higher image quality provided by DSA is not clinically needed.

to undergo a poster anterior cardiac catheter ablation procedure during the first trimester, fluoroscopic imaging with an empty bladder delivers the lowest absorbed dose the conceptus (ICRP 2013, Birnie *et al* 2011, Damilakis *et al* 2005). Alternatively, for AP projections, a full bladder will decrease dose by pushing the uterus in the PA direction. If the procedure is not an emergency, an estimate of potential conceptus dose should be obtained before the procedure is performed, and this information should be utilized in planning optimization for performing the actual procedure.

Pre- and post-procedure related dosimetric considerations
When a high-dose procedure is performed and when the conceptus is estimated to be in the primary or close to the X-ray beam, all technical factors should be recorded to allow subsequent fetal dose estimation. In addition, appropriate dosimetry should be employed to manifest measurements of entrance surface dose at several locations on the pregnant patient. Multiple dosimeters, placed in several locations on the patient (anterior and posterior to the uterus), will aid in the development of post-procedure dose estimates.

Radiation dose monitoring should be assigned to a specific individual depending on the institution's policies and needs. Dose monitoring ensures that the operator is aware of how much radiation is being administered, since it is usual for the operator to concentrate on the interventional procedure and they may lose awareness of the patient's radiation dose. It is the operator's responsibility to be informed about dose levels and to include radiation dose in the continuous risk–benefit equilibrium of the procedure. Dosimetric information should be recorded in the patient's medical record as soon as is practical after the completion of the procedure. In addition to any conceptus dose estimates, this should include all of the following that are available from: peak skin dose, cumulative kerma at the interventional reference point, air kerma product, fluoroscopy time and number of fluorographic images acquired.

The evaluation of conceptus doses from abdominal fluoroscopy is challenging and subject to greater uncertainty. With fluoroscopy and angiography, the X-ray tube position relative to the patient may change several times throughout the examination. In addition, radiation is not used continuously, but is employed sporadically at different times during the study. The exact parameters are almost never known, and conceptus dose estimates have often been based on a 'typical' study in the literature. These estimates, along with an assessment of the uncertainty range, can be expressed to the interventionalist and the referring clinician. Although prospective dose estimation with the use of TLD dosimeters placed in several locations on the patient (anterior and posterior to the uterus) to gather measurements of entrance surface dose at several locations on the pregnant patient, it is typically performed less often than retrospective dose estimation. Retrospective dose estimation can initially be made using tables of available estimates. In either case, if the initial estimation of dose is 10 mGy or greater a more detailed dosimetry estimate should be developed. All such dose estimates should be properly documented and included as part of the patient's medical record.

Direct measurement models in phantoms have been performed for various diagnostic examinations, and investigators have measured uterine depth dose within a humanoid phantom for various kVp beams of diagnostic quality. Using tables of such dose measurements, conceptus doses could be estimated from the knowledge of the conceptus localization (e.g. using sonography) and the beam parameters used in the procedure.

Computational anatomic models have also been developed for conceptus dose estimation. Such models can be stylized models that comprise organs described by mathematical equations, or models that contain voxelized tomographic data from segmented computed tomography (CT) sequences. Recently, advanced simulation phantoms have also been proposed that not only use constructive solid geometry but boundary representations that can be computationally adaptable. Normalized conceptus doses for abdominal radiographic examinations have been estimated by using phantom models with Monte Carlo methods that use various radiation codes such as the Monte Carlo N-particle (MCNP) transport code or Monte Carlo software PCXMC from Radiation and Nuclear Safety Authority of Finland (STUK) software developed specifically for diagnostic X-ray projections. Typically, these methods use beam characteristics, such as kVp, total filtration values, and field size, to estimate organ or conceptus doses with results that can agree with reported published or measured dose data within approximately 20%–40% (Schultz *et al* 2003).

6.2 CT during pregnancy: methods for dose optimization

6.2.1 Introduction

The number of pregnant patients subjected to CT imaging has been steadily increased over the last few decades following the rapid evolution of multidetector CT technology and increased availability (Lazarus 2009, Goldberg-Stein 2011). Optimization of CT exposures on pregnant patients aims for the reduction of the radiation burden to both the expectant mother and the conceptus. CT examinations performed on pregnant patients may be classified in two broad categories. Type A CT examinations: the developing embryo/fetus is partly or completely included in the primarily exposed body region during acquisition, and type B examinations: the developing embryo/fetus lays outside the primarily exposed body region during acquisition. Examinations of type A may result in dose to embryo/fetus tissues typically in the range 10–40 mGy with the magnitude of dose depending mainly on the CT exposure settings and dose sparing tools/techniques employed during acquisition (ICRP 2000). Examinations of type B may result in dose to embryo/fetus well below 1 mSv and the magnitude of dose is mostly determined by the distance of the embryo/fetus tissues from the boundary of the primarily exposed body region and secondarily by exposure settings and dose sparing tools/techniques employed during acquisition. The same type of CT examination may be categorized as type A or B depending on the specific patient. For example, an upper abdomen CT examination on a pregnant patient at the first trimester is type B, whereas the same CT examination on a pregnant patient at the end of third trimester may be categorized in type A. CT examinations of either type should be optimized in order to reduce the expected radiation burden to both the expectant mother and embryo but apparently this is particularly important in type A CT examinations.

Optimization of CT exposure on a pregnant patient is a difficult task that requires familiarization with the specific technological features of the equipment used and collaboration of the radiologist, the medical physicist and the operator. Some general guidelines and practical hints are discussed here.

6.2.2 Optimizing CT examinations of pregnant patients: recommendations for radiologists

Consideration of alternative imaging modalities

Radiologists should strictly adhere to 'justification principle' when a pregnant patient is referred for CT examination. Alternative imaging modalities should be considered and discussed with the referring physician and the medical physics expert taking into account the clinical urgency, the expected level of radiation risk and the availability of other imaging modalities (Goldberg-Stein 2011). Radiologists should refuse performing a type A CT examination on a pregnant patient if it is not adequately justified or there are other imaging modalities that may successfully confront the specific diagnostic task.

Modification of standard CT imaging acquisition protocols

In the case of a pregnant patient referred for a justified CT examination, the standardized CT imaging acquisition protocol for adult females should be modified towards reducing the risk for both the examined expectant mother and the embryo/fetus. Apart from the optimization of the exposure parameters and dose sparing tools/techniques discussed later in this chapter, there are additional means to reduce the radiation burden to the expectant mother and embryo/fetus. If a pregnant patient is referred for a multiphase CT examination including both unenhanced and contrast-enhanced CT imaging phases, the radiologist should consider reducing the number of acquired imaging phases provided that the clinical information expected from the examination may be still derived with confidence. Omission of the contrast-enhanced CT imaging phase may also prevent the risks associated with the administration of contrast medium, on top of reducing the total radiation burden to both the expectant mother and embryo/fetus. It is noted that, apart from the risk for induction of allergic reaction to the expectant mother, administration of iodinated contrast medium during pregnancy has been associated with increased risk of hypothyroidism in the developing embryo (Tirada *et al* 2015). Also, if the CT acquisition protocol involves the use of automatic mA modulation, the radiologist should consider setting the desired image quality level slightly lower than in common adult CT examinations. This would allow the mA modulation system to achieve extra reduction of the exposure parameters during acquisition. High experience and familiarization with the specific diagnostic task are prerequisite for the radiologist to efficiently minimize the number of CT acquisitions on a pregnant patient and appropriately set the desired image quality of the resulting CT image series. Finally, the radiologist should always encourage operators to imperatively and meticulously use any dose sparing tool/technique available whenever a pregnant patient is to be subjected to CT imaging.

6.2.3 Reducing conceptus dose from CT examinations performed on the expectant mother: recommendations for operators

Setting the primarily exposed body region extent to minimum

In sequential scan acquisition mode, the scanning length is defined by the prescribed image volume as defined by the operator on the scout view(s) acquired prior to the CT scan. In helical scanning, the scanning length is expanded by a certain extent beyond the boundaries of the prescribed image volume due to z-overscanning. In either acquisition mode, the longer the prescribed image volume along the z-axis the longer the scanning length. Operators should always meticulously set the boundaries of image volume to include only the body region under investigation. Thus, the primarily exposed body region is minimized. Setting the image volume length to minimum is a very efficient means to reduce the dose to both the expectant mother and the embryo/fetus. It is simple and very efficient, and it does not affect the image quality of the produced image series. Reducing the image volume length by 1–3 cm may result in 24% and 56% radiation dose reduction to embryo/fetus from chest and upper abdomen CT examinations during the first trimester (Ryckx 2018). In the same context, sequential acquisition mode should be preferred over helical acquisition in CT examinations of pregnant patients whenever the clinical diagnostic task may be successfully confronted either way. For a specific prescribed image volume, the primarily exposed body region is much less when sequential instead of helical acquisition mode is employed due to the absence of overscanning in sequential scanning. Whenever helical acquisition is clinically demanded, adaptive section collimation should be employed, if available, to reduce the overscanning extent (Deak *et al* 2009).

Setting kV and mAs low

For several decades, adult CT examinations were commonly performed at the same tube voltage value, that is 120 kV. Only the mAs value was commonly modified by the operator depending on patient body size to reduce the radiation dose to slim patients; but still the reduction of the mAs value, below that prescribed in the standard CT imaging protocol, was rather empirical (Tsapaki *et al* 2010). Simplified charts and formulas were proposed to achieve the appropriate mA selection depending on the size of the patient to be examined. During the last decade, tube voltage was identified as a crucial parameter towards reducing patient radiation dose from CT and the use of lower than 120 kV tube voltages were highly recommended especially for small size adults and paediatric patients. The use of 100 kV and or even 80 kV may occasionally be the most appropriate choice for pregnant patients referred for CT examination. However, reducing the tube voltage to optimum requires a high level of experience by the operator since it should be accompanied by an appropriate increase of mAs setting, if adequate image quality is to be reserved in the resulting image series.

The introduction of automatic mA modulation mechanisms, and later automatic selection of kV in combination with automatic mA modulation systems, aimed to achieve optimization of CT exposure parameters without the requirement of highly experienced operators able to empirically reach a 'wise' kV/mAs selection.

In the case of a pregnant patient referred for a CT examination that does not involve direct exposure of the embryo/fetus tissues (e.g. head and chest CT at any trimester or upper abdomen CT in first trimester), automatic exposure control (AEC) systems should be always applied, if available. Thus, overexposure or low image quality in the resulting image series may be avoided. The desired image quality of the resulting CT image series may be set at the same level as in non-pregnant adult females, since forcing AEC to reduce the expected minor dose to conceptus is not crucial.

In the case of a pregnant patient referred for an examination that does involve direct exposure of the embryo/fetus tissues (e.g. abdomen and/or pelvis CT at any trimester), the use of automatic kV/mA adaptation schemes should be applied with great caution (Goldberg-Stein *et al* 2012). During early pregnancy, the size of a patient's abdomen has not yet increased and therefore application of automatic exposure control systems may efficiently minimize the radiation burden to both the expectant mother and the embryo/fetus. Moreover, setting the prescribed image quality of the resulting CT image series slightly lower than non-pregnant adults of the same size should be considered to further reduce the embryo/fetus dose. Besides, during late pregnancy the abdominal size of the expectant mother is significantly increased. Therefore, the use of automatic kV/mA adaptation may result in a highly elevated dose for the embryo/fetus especially when the desired image quality is maintained as in non-pregnant females of the same size. In general, optimization of the automatic kV selection/mA modulation scheme in patients at late pregnancy referred for abdomen or pelvis CT is cumbersome. It requires high familiarization with the behavior of the available automatic exposure system when the modulation strength and/or the mimimum/maximum mA are altered.

Organ-based mA modulation has been proposed to reduce radiation dose absorbed by radiosensitive tissues that lie superficially in the anterior surface of the patient. There are no scientific data in the literature supporting the use of organ dose modulation in pregnant patients referred for abdomen/pelvis CT as a means to reduce embryo/fetus dose. During early pregnancy, the embryo is located rather centrally in the body of the expectant mother, while fetal tissues during late pregnancy may be found centrally or even posteriorly. Therefore, application of organ-based tube current modulation may occasionally end up with the opposite than expected result.

In the case of pregnant patients referred for cardiac CT, prospective ECG-gated acquisition mode should be always preferred over retrospective acquisition. If retrospective acquisition mode is clinically demanded, ECG-based mA modulation should be employed. The use of ECG-mA modulation may result in considerable absolute reduction of embryo dose from cardiac CT especially in the case of late pregnancy where fetal tissues may be found in close proximity to a primarily exposed body region.

Appropriate setting of beam collimation, focal spot size and pitch
Beam collimation is defined by the number of active detector elements (data channels) times the width of each detector element. For example, a 64-slice CT system with minimum detector element width of 0.625 mm allows acquisitions with beam collimation of (a) 64 × 0.625 mm, (b) 32 × 1.25 mm, and (c) 32 × 0.625 mm, among others. The beam width at the isocenter is 40 mm for beam collimations (a) and (b) and 20 mm for (c). In general, collimations of large beam width (i.e. (a) and (b) in the above example) are associated with higher geometric efficiency and may be considered preferable. Indeed, $CTDI_{vol}$ for a wide beam is lower than the corresponding value of a narrow beam when all other exposure parameters (i.e. kV, mAs, pitch, etc) are held constant (Perisinakis *et al* 2009). Also, the use of a small focal spot is strongly recommended given that it is associated with less penumbra and therefore superior geometric efficiency for a certain beam collimation (Perisinakis *et al* 2009). In most systems, the size of the focal spot is automatically changed from small to large when the mAs setting exceeds a certain threshold. Therefore, operators should consider setting the mAs value at a level where acquisition with a small focal spot is allowed. Also, in CT acquisition with mA modulation, setting the maximum mA value below this threshold would allow acquisition with a small focal spot.

The active detector element width defines the minimum reconstructed slice thickness. In general, thin reconstructed slice thickness requires higher exposure settings to maintain noise. Therefore, selection of higher reconstructed slice thickness allows for the use of lower exposure parameters. The recommendation for 'scan thin-view thick' refers to the use of a thin detector element width and thick reconstructed slice width. For example, the use of 64 × 0.625 mm beam collimation may allow for reconstruction of 1.25 or 2.5 mm images with acceptable noise, while the thin 0.625 mm images, which are expected to have relatively higher noise, remain as an option and may be used for volumetric/isotropic 3-d imaging.

Since the amount of dose received by patient tissues is inversely proportional to pitch, the use of pitch <1 in either sequential or helical acquisitions should be avoided in pregnant patients especially for CT examinations of the abdomen and pelvis. Besides, pitch values much higher than one should be used with caution since by increasing the pitch value the overscanning extent may be increased by a few cm (Tzedakis *et al* 2005). Expanding the primarily exposed body region might result in approaching the scanned region boundary to the embryo/fetus or even primary exposure of embryo/fetus tissues that would not have been exposed primarily if pitch one was used.

Use of iterative reconstruction (IR) algorithms
Modern CT suites provide very efficient IR tools that have been reported to achieve much higher image quality at reduced patient dose levels in comparison to the standard filtered back-projection reconstruction algorithm (Geyer 2015). Whenever available, IR should be employed in CT examinations of pregnant patients, since substantial dose savings for the embryo/fetus may be achieved. Dose savings well exceeding 50% have been reported when IR is used in average-and large-size

patients. Since the abdominal size of pregnant females is considerably elevated during the second and third trimester, dose reduction of this level could be also achieved in abdomen/pelvis CT examinations of pregnant patients (Imai *et al* 2017).

Shielding the conceptus

In the case of pregnant patients referred for CT, the use of a radioprotective apron to wrap the patient's abdomen has been proposed as a means to reduce embryo/fetus dose. This is an efficient dose sparing tool in head, neck or thorax CT scans where the scanned body region is far from the lower abdominal area containing the conceptus. Wrapping the abdominal area with radioprotective garments may achieve a remarkable reduction of the external scatter component but not the internal scatter. The internal scatter component is minor when the scan is located far from the embryo tissues and therefore, a considerable % embryo/fetus dose reduction may be achieved by shielding the conceptus (Iball 2008, Chatterson *et al* 2014). The absolute gain, however, is minor since the expected dose to the embryo/fetus is minor. When the embryo tissues are in close proximity to or partly included in the primarily exposed body region, the use of radioprotective garments to wrap the patient may be associated with the opposite than desired effect when mA modulation is employed (Ryckx 2018). In this case, the internal scatter component may be much higher than the external scatter component. The automatic exposure mechanism may significantly elevate exposure settings if the CT beam falls on the radioprotective garment thus increasing internal scatter. Therefore, shielding the abdominal area of a pregnant patient should be avoided when the image volume is in close proximity with the embryo/fetus location.

References

Birnie D, Healey J S and Krahn A D *et al* 2011 Prevalence and risk factors for cervical and lumbar spondylosis in interventional electrophysiologists *J. Cardiovasc. Electrophysiol.* **22** 957–60

Chatterson L C, Leswick D A, Fladeland D A, Hunt M M, Webster S and Lim H 2014 Fetal shielding combined with state of the art CT dose reduction strategies during maternal chest CT *Eur. J. Radiol.* **83** 1199–204

Damilakis J, Perisinakis K and Theocharopoulos N *et al* 2005 Anticipation of radiation dose to the conceptus from occupational exposure of pregnant staff during fluoroscopically guided electrophysiological procedures *J. Cardiovasc. Electrophysiol.* **16** 773–80

Deak P D, Langner O, Lell M and Kalender W A 2009 Effects of adaptive section collimation on patient radiation dose in multisection spiral CT *Radiology* **252** 140–7

Geyer L L, Schoepf U J and Meinel F G *et al* 2015 State of the art: iterative CT reconstruction techniques *Radiology* **276** 339–57

Goldberg-Stein S, Liu B and Hahn P F *et al* 2011 Body CT during pregnancy: utilization trends, examination indications, and fetal radiation doses *Am. J Roentgenol.* **196** 146–51

Goldberg-Stein S A, Liu B, Hahn P F and Lee S I 2012 Radiation dose management: part 2, estimating fetal radiation risk from CT during pregnancy *Am. J. Roentgenol.* **198** W352–6

Iball G R, Kennedy E V and Brettle D S 2008 Modelling the effect of lead and other materials for shielding of the fetus in CT pulmonary angiography *Br. J. Radiol.* **81** 499–503

ICRP 2000 Publication No. 84 Pregnancy and medical radiation. *Ann. ICRP* **30** 1–43

ICRP Publication 103 2007 International commission on radiological protection. The 2007 recommendations of the international commission on radiological protection *Ann. ICRP* **37** 2–4

ICRP Publication 120 2013 Radiological protection in cardiology *Ann. ICRP* **42** 1–125

Imai R, Miyazaki O, Horiuchi T, Asano K, Nishimura G, Sago H and Nosaka S 2017 Ultra-low-dose fetal CT with model-based iterative reconstruction: a prospective pilot study *Am. J. Roentgenol.* **208** 1365–72

Lazarus E, Debenedectis C and North D *et al* 2009 Utilization of imaging in pregnant patients: 10-year review of 5270 examinations in 3285 patients 1997–2006 *Radiology* **251** 517–24

Perisinakis K, Papadakis A and Damilakis J 2009 The effect of x-ray beam quality and geometry on radiation utilization efficiency in multi-detector CT imaging *Med. Phys.* **36** 1258–66

Ryckx N, Sans-Merce M and Schmidt S *et al* 2018 The use of out-of-plane high Z patient shielding for fetal dose reduction in computed tomography: literature review and comparison with Monte-Carlo calculations of an alternative optimisation technique *Physica Med.* **48** 156–61

Schultz F W, Geleijns J, Spoelstra F M and Zoetelief J 2003 Monte Carlo calculations for assessment of radiation dose to patients with congenital heart defects and to staff during cardiac catheterizations *Br. J. Radiol.* **76** 638–47

Tirada N, Dreizin D, Khati N J, Akin E A and Zeman R K 2015 Imaging pregnant and lactating patients *Radiographics* **35** 1751–65

Tsapaki V, Rehani M and Saini S 2010 Radiation safety in abdominal computed tomography *Semin. Ultrasound CT MR* **31** 29–38

Tzedakis A, Damilakis J, Perisinakis K, Stratakis J and Gourtsoyiannis N 2005 The effect of z-overscanning on patient effective dose from multi-detector spiral computed tomography examinations *Med. Phys.* **32** 1621–9

Chapter 7

Optimization of examinations performed on paediatric patients

Antonios Papadakis, John Stratakis and Virginia Tsapaki

Radiation risk to human health has been the subject of abundant research and debate. Attention has been emphasized on children as they are often considered to be more vulnerable to radiation. Justification and optimization are the fundamental principles of radiation protection in medical exposures. Paediatric radiography and fluoroscopy procedures may require dedicated anatomy-specific examination protocols. In addition, correct positioning is important in radiographic/fluoroscopic image quality and will also affect the radiation dose in computed tomography (CT) examinations. Many studies have reported large dose reductions using modified CT exposure parameters. However, the wide range reported, as well as the different CT applications reveal the difficulty in standardizing CT procedures. In the following sections, practical ways of optimization methods will be presented.

7.1 Optimization of radiographic and fluoroscopic examinations performed on paediatric patients

7.1.1 Introduction

Exposure to low radiation doses such as those delivered to patients during diagnostic procedures may hold a probability, even if small, of inducing cancer years or even decades after the examination (UNSCEAR 2006). The benefits for patients overshadow the radiation risks when these procedures are appropriately prescribed and performed. Indeed, for some tumor types, the paediatric population is more sensitive to radiation exposure than adults. This increased sensitivity varies with age, with the younger ages being more at risk (UNSCEAR 2013a, 2013b). Studies have also shown that radiogenic tumor occurrence in children is more inconstant than in adults and depends on the type of tumor, and on the child's sex and age at

doi:10.1088/978-0-7503-1317-9ch7

irradiation. These studies on the differences in radio sensitivity between children and adults have found that children are more sensitive for the development of thyroid, brain, skin and breast cancer and leukemia. The susceptibility of children to radiation-induced cancer has been of interest for over half a century. Cancers related to childhood irradiation on average result in more years of life lost than those related to exposure in adulthood. Children have a longer life expectancy resulting in a larger timeframe for demonstrating long-term radiation-induced effects. Radiation-induced cancer may have a long latency period that varies with the type of cancer and the dose received. The latency period for childhood leukemia is generally less than 5 years, while this period for some solid tumors is usually decades.

Guidelines to manage and reduce radiation dose from X-ray examinations while maintaining diagnostic image quality have been documented (NCRP Report No. 68 1981 and EUR16261 1996a). These guidelines have been mostly based on practices relating to conventional screen-film radiography. Conventional radiography has evolved to computed radiography (CR) and more recently to digital radiography (DR). The radiology community admits that there might be a potential gap in the knowledge for those professionals that were experienced in the analog screen-film technology and are currently asked to work with the new digital technology. This chapter focuses on the practices and recommendations towards the optimization of DR and fluoroscopy examinations performed in paediatric patients.

7.1.2 Radiation protection concepts

Justification and optimization are the fundamental principles of radiation protection in medical exposures. Although the radiation protection system is also based on dose limitation, in the case of medical exposures dose limits are not applied because they may reduce the effectiveness of the patient's diagnosis or treatment, thereby doing more harm than good (ICRP 2007). Medical exposures shall be justified by weighing the expected diagnostic or therapeutic benefits against the potential radiation damage, considering the benefits and the risks of alternative techniques that do not involve exposure to radiation. The principle of justification applies whether the application of a specified procedure to an individual patient is judged to do better than harm to the patient. The responsibility of justifying a procedure for a patient falls upon professionals directly involved in the healthcare delivery process (e.g. referrers, radiologists). When indicated and available, imaging media that do not use ionizing radiation are preferred, especially in children and in pregnant women (particularly when direct fetal exposure may occur during abdominal/pelvic imaging). The possibility of deferring imaging to a later time when the patient's condition may change also must be considered. The final decision may also be influenced by cost and availability of resources. Imaging referral guidelines help healthcare professionals make decisions by providing clinical decision-making tools created from evidence-based criteria. In the context of radiation protection, optimization denotes keeping doses 'as low as reasonably achievable' (ALARA). For medical imaging, ALARA entails delivering the lowest possible dose necessary to acquire adequate diagnostic data images: best described as 'managing the radiation dose to be in balance with the medical purpose'.

7.1.3 Appropriateness and clinical decisions

The most effective means to decrease radiation dose associated in paediatric imaging is to reduce unnecessary or inappropriate procedures. Justification of a procedure by the referring clinician who is responsible for the management of the patient and the radiologist who selects the most appropriate imagine technique, is a key gauge to evade needless irradiation before a patient undergoes imaging. Overuse of diagnostic radiation triggers avoidable risks and can add to health expenditure. In some countries, a considerable fraction of radiologic examinations are of arguable merit and may not provide a benefit to patient healthcare (Hadley 2006, Oikarinen *et al* 2009). The real magnitude of unjustified risk resulting from inappropriate use of radiation in paediatric imaging remains ambiguous. Possible reasons for inappropriate radiation procedures in children may include low understanding of radiation doses and associated risks, incorrect clinical information provided for justification, concern about malpractice litigation (defensive medicine), lack of discussion between referrers and radiologists, reliance on personal or anecdotal experience not supported by evidence-based medicine and lack of availability of alternate imaging resources-expertize and/or equipment. Duplication of imaging already performed at other healthcare facilities constitutes a significant fraction of unnecessary examinations. Previous investigations should be recorded in detail and be accessible to other healthcare providers that is at the point of care. However, intricacies and advances in medical imaging may render it difficult for referrers to follow changes in evidence-based standards of care Guidance for justification of imaging is usually provided by professional societies in conjunction with national departments of health to produce recommendations based upon the best available evidence, including expert advice, designed to guide referrers in appropriate patient management by selecting the most suitable procedure for clinical indications. Referral guidelines for appropriate use of imaging provide information on which imaging exam is most apt to yield the most informative results for a clinical condition, and whether another lower-dose modality is equally or potentially more effective. Although referral guidelines are not mandatory, a referrer should have good reasons to deviate from these recommendations. Greater and more effective communication between referrers and radiologic medical practitioners would facilitate the optimization process. Information provided by the referrer (i.e. legible and clearly expressed requests) should include the clinical questions to be addressed by the imaging procedure. Through clinical audit, medical procedures including medical imaging should be systematically reviewed against established standards for good medical practice. This aims to improve the quality and the outcome of patient care, thus also contribute to advancing radiation safety culture.

7.1.4 Assessment of radiation dose and image quality—a primer

There are several quantities that are used for evaluating doses to patients. The dose quantities that can be measured for radiographic and fluoroscopic exposures are the entrance-surface dose (ESD) and the dose-area product (DAP). The ESD is the dose to the skin at the point where X-rays enter the body and includes both the incident air kerma and radiation backscattered from the tissue. It can be measured with small dosimeters placed on the skin or calculated from radiographic exposure factors of tube output. The DAP is the product of the dose in air (air kerma) within the X-ray beam and the beam area and is a quantity of all radiation that enters a patient. It can be measured using an ionization chamber fitted to the tube. DAP and ESD can be used to monitor, audit and evaluate radiation doses from a wide variety of radiological examinations. Diagnostic reference levels (DRLs) for examinations have been established in terms of the ESD or DAP. Effective dose is useful for comparing doses from different types of examination in general terms for a reference patient and assessing changes in the dose for a reference patient during the process of optimization. Another simple way to derive quantity that can be equally practical, is the energy imparted to the body by an X-ray exposure (Chapple *et al* 1994). This includes all the energy absorbed from an X-ray beam and provides a representation of the relative harm than a measurement of ESD and does not include the complications and approximations involved in the calculation of effective dose.

Image quality can be quantified in terms of the characteristics; contrast, sharpness (or resolution) and noise. Contrast is an outcome of the different attenuation of X-rays in tissue, sharpness is the aptitude to display small details, and noise refers to the random fluctuations across the image that tend to recondite the detail. Evaluation and diagnosis from the image require structures of interest to be differentiated from the background. The difference between the film optical density of a structure of interest and that of the background can be thought of as the signal. Random fluctuations across the image or film can occur, which are superimposed. These are referred to as noise, and result from several causes; quantum mottle due to statistical variations because of a limited number of photons, the granularity or finite grain size of the film, and anatomic dissimilarities in structure density through the tissue. The fluctuations affect the detection of low-contrast structures. An optimized radiograph should be limited by quantum mottle. Objective methods of evaluating image quality quantify the imaging performance in terms of the signal reproduction for details of different sizes, using quantities such as the modulation transfer function (MTF), and their visibility within the noise generated by the imaging system, such as the detective quantum efficiency (ICRU Report 54 1996, Martin *et al* 1999).

7.1.5 Practices for optimization of radiographic examinations performed on paediatric patients

Employment of exposure technique charts

Paediatric radiography and fluoroscopy procedures require dedicated anatomy-specific examination protocols. These protocols should include preprogrammed exposure technique charts, which vary from infants to obese adolescents. The anatomy-specific examination protocol presets provided by the manufacturer may not be appropriate for children. The medical imaging physicist, radiologist, radiologic technologists and manufacturer should work together to develop exposure technique charts for paediatric patients that are programmed into the unit's preset examination protocols. The employment of the anatomy-specific exposure technique charts eliminates much of the concern regarding the selection of different variables such as kVp, mA, grid use and tube-to-detector or tube-to-patient distance. The charts enable radiologists and technologists to determine the radiation exposure level that provides images of diagnostic quality on the basis of the ALARA principle. A complete exposure technique chart should include at least the following variables for each X-ray unit: mA, exposure time, kVp, source-to-image receptor distance, focal spot size, use of grid and the grid ratio, automatic exposure control (AEC) detectors, and range of acceptable exposure index (EI).

By measuring patients' body size with calipers will ensure that a standardized technique is selected. Knowing the thickness of the body anatomy examined, one can tailor the kVp, filtration, and mAs for the study of the individual child. The objective is reproducible and consistent images for children with body anatomies of the same size. Furthermore, technologists must carefully select the optimum kVp for the paediatric patient's anatomy under investigation. Selection of the optimum kVp is more critical in examinations of infants and young children because their tissue composition displays less subject contrast. Infants and children have bones with less calcification than adults, which requires a lower kVp compared. kVp may thus be reduced to form an image that is still of diagnostic value. Exposure technique charts should be monitored and revised continuously to ensure exposure techniques are producing diagnostic images within the ALARA principle. A best practice in DR is to use exposure technique charts that are continuously improved and applicable to a wide range of patient sizes.

Radiologic technologists who use AEC settings for imaging paediatric patients should follow the Image Gently digital safety checklist, which emphasizes that radiographers must be diligent in ensuring that the appropriate kVp, exposure time and image detector have been selected. Manual technique selection may be needed in paediatric radiography when the anatomical region being imaged is smaller than the size of the ionization chamber.

The advancement of technology to digital radiology

DR has largely replaced screen-film radiography in radiology departments worldwide. Radiologists, technologists and medical physicists are responsible for in-depth understanding and proper use of DR equipment. DR systems image at latitude that is approximately 100 times wider than in screen-film radiography. A wider range of patient exposures may be used given that technologists can compensate effectively an underexposed or an overexposed image by adjusting brightness, contrast or window display values through image processing (Butler 2010). Underexposed images may be realized as noisy with a high quantum mottle. In contrast, overexposed images contain a low level of noise and they are attractive to radiologists (figure 7.1). The latter images may go unnoticed with this practice contributing to the increase of patient collective dose. Image acquisition and processing may vary considerably among different equipment produced by different radiography vendors. The techniques required to produce digital images of acceptable quality differ from those used in screen-film radiography or even in CR. In DR, digital detectors of different technologies may need different techniques in image processing due to differences in the detective quantum efficiency (DQE) of the employed detector materials. In radiology departments with imaging units from different vendors and technologies, confusion may arise on the exposure techniques required per examination. Radiologists, technologists and medical physicists should adopt a routine approach to produce diagnostic radiographs based on the continuing review of the EI and the assessment of quality of the obtained images on an individual basis. The detectors used in DR allow imaging in a much wider dynamic range compared to that in conventional screen-film radiography. Digital detectors can respond linearly to very low levels as well as very high levels of radiation (i.e. 100 μR to 100 mR). The wide dynamic range combined with the capabilities allowed by the digital image processing techniques results in images that cannot be generated with conventional screen-film systems.

The use of IEC standard for exposure indicator (EI)

The exposure indicator (EI) provides feedback to the technologist regarding the proper radiographic exposure for a specific examination by demonstrating the relative exposure at the image detector based on its efficiency and sensitivity to the incident X-ray photons. The variation in the way EI was used to be determined

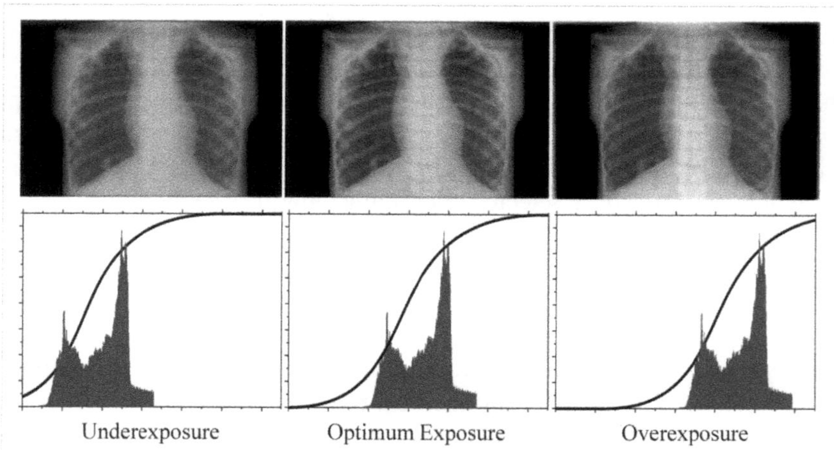

Figure 7.1. AP digital chest radiographs of a 5-year-old anthropomorphic phantom under different exposure conditions (under exposure, optimum exposure and over exposure) using an AGFA DR radiography unit (DX-D600, AGFA Healthcare, Belgium). The image processing capabilities enable the production of diagnostic images at three exposure conditions. Exposure parameters: 80 kVp, 0.4 mAs (under exposure), 1.2 mAs (optimum exposure), 8 mAs (over exposure), source-to-image distance 110 cm, additional filtration: 1 mm Al + 0.1 mm Cu, AEC deactivated.

and reported (terms, units, calibration conditions, etc) among different manufacturers of DR equipment has forced the International Electrotechnical Commission (IEC) to establish an EI terminology standard in 2008 in accordance to the recommendations of Image Gently Paediatric Digital Radiography Summit (Don 2011) and the American Association of Physicists in Medicine (Shepard 2009, IEC 2008).

The new IEC standard has introduced three terms: (i) target exposure index (EI_T), (ii) EI, and (iii) deviation index (DI) (Don 2012).

(i) The EI_T represents the ideal exposure at the image detector for each examination when an image is optimally generated. The EI_T values may differ for each anatomical region and projection orientation, and may vary on the sensitivity of the detector of the imaging unit. The EI_T is set either by the manufacturer or by the user for each anatomy-specific examination and imaging unit such as determined by statistical averaging of a series of at least 50 exposures using a typical phantom or patient's group having the same anatomical characteristics. The EI_T should indicate the exposure level that generates images of acceptable quality to radiologists. Few data are available about establishing reference EI_T values. A single value is needed for every examined anatomical region. It should be noted that similar anatomical regions such as abdomen and pelvis, lower legs and upper arms or knee and shoulder may share similar EI_T values provided that images are acquired at similar kVp and that image processing algorithms for display are the same.

(ii) The EI constitutes a measure of radiation exposure at the image detector in a specified region of the image. The EI is linearly associated to mAs. The EI is derived from the signal-to-noise ratio (SNR) measured on the image upon exposure (Kallman *et al* 2011). If a thick or dense body part is imaged then the exposure factors need to be adjusted to compensate for the increased beam attenuation. Correspondingly, when a thin or a paediatric patient is imaged, the exposure factors need to be reduced so that the same number of photons reaches the detector to achieve the same SNR. This process results in the same EI at the lower dose for the patient. It should be noted that EI is not a measure of patient dose rather it simply indicates the detector's exposure. The EI depends on several parameters such as the examined anatomical region, its thickness, tube-to-detector distance, kVp, added filtration, presence of artifacts, collimation and the type of employed detector (Seibert and Morin 2011). At the same patient, kVp and beam filtration, doubling the mAs will double the EI. When kVp changes and mAs is adjusted to maintain a constant exposure, the EI changes non-linearly. Due to differences in equipment, exposure factors and the radiologist's subjective image quality preferences, EI may not demonstrate variations in radiation exposure among different institutions. The EI can be safely used for comparison between exposures for the same examination type on the same imaging unit.

(iii) The DI indicates how far the actual EI value deviates from the examination specific EI_T based on the following scheme: $DI = 10 \times Log_{10}(EI/EI_T)$

(ACR-AAPM-SIIM 2013, Cohen *et al* 2011). When EI is higher than EI_T, the DI is positive, while when EI is lower than the EI_T, the DI is negative. The DI ranges from negative to positive values in under to overexposures, respectively. In an ideal exposure, when EI equals EI_T, DI is zero. One deviation unit denotes that exposure was performed at 20% less (−1) or 26% more (+1) radiation dose relative to the ideal (0). Moreover, a DI of ±3 indicates that exposure is half or double relative to EI_T. The technologist must specify the correct anatomical region and radiographic projection so that the correct EI_T value is retrieved for the calculation of the corresponding DI value. The DI provides a direct feedback to the technologist of the appropriateness of the performed radiographic exposure. The technologist can refer to the calculated metrics to realize what changes should be performed in the exposure factors in the upcoming examinations. The employment of DI is important in the effort to optimize radiographic studies especially in children. Each radiology department should ensure that DI is displayed on each image viewed by the PACS system, so that radiologists can detect inappropriate exposures. It should be noted that acceptable DI values may not always facilitate good image quality (Shepard 2009). The DI is simply a number that demonstrates how close the performed exposure was to the ideal. Each radiograph should still be evaluated on an individual basis for its diagnostic value and exposures should not be repeated solely on an inappropriate EI since this will contribute to unnecessary patient exposure.

It is important to emphasize that EI is an indicator of the amount of radiation incident on the detector and not the radiation dose to the patient. Technologists must understand the relationship between the exposure technique factors used and the resulting EI values. In digital processing of the acquired image, identification of exposure field borders is involved. Inaccurate exposure field recognition may cause errors in standard deviation readings required for EI calculation. Another limitation of the EI is that different manufacturers use varying methods to determine the image regions required to analyze and generate an EI value. Moreover, the wide exposure range offered by digital detectors and issues such as poor collimation, patient positioning and patient's body habitus, can cause EI to be higher or lower than anticipated. A performed examination with an acceptable EI should always be accompanied with the visual evaluation of its diagnostic value. Since EI has limitations, technologists in collaboration with the radiologists need to carefully assess whether a repeat of the examination is required.

Selection of kVp

The best practice in DR for children is to use the highest kVp within the optimal range for the position and part coupled with the lowest amount of mAs needed to provide an adequate exposure to the image receptor (Herrmann *et al* 2012). For body exposures, a high kVp results in lower patient attenuation, and therefore lower dose for the same detector exposure. The optimal kVp is usually higher for bone to soft tissue contrast and for thicker objects (Seibert 2004). The kVp should be increased with ascending age/size group due to increase in tissue thickness that requires more photon penetration. Several studies have been performed on the optimization of kVp, but many of these have not been paediatric specific (Herrmann *et al* 2012, Hess *et al* 2011, Schaefer-Prokop *et al* 2008).

For distal extremity exposures, optimization of kVp should be based on lowering the kVp and increasing the mAs to achieve a constant patient dose instead of a constant detector dose but at an increased contrast-to-noise ratio (CNR) (Hess *et al* 2011). In such a case the detector exposure decreases resulting in a lower EI. Distal extremity kVp should vary between 40 kVp for a baby hand of 1 cm thickness, up to 66 kVp for an adult size knee of 12 cm thickness. Uffmann *et al* have also investigated the same principle of lowering kVp and increasing mAs to achieve a constant patient dose instead of constant detector exposure in adult chest radiography (Uffmann *et al* 2005). Carestream Health have also written a white paper, with a basic overview of maximizing dose efficiency for paediatric imaging through optimization of kVp (Carestream Health 2012).

Table 7.1 shows the DAP (cGycm^2) at varying kVp and mAs for a typical DR system (AGFA DX-D600, AGFA Healthcare, Belgium). As collimation (30 × 20 cm) and tube-to-image distance (100 cm) were constant in these measurements, DAP may be considered directly proportional to the entrance skin dose for a body part of the same thickness. 40 kVp 4 mAs^{-1} has a similar DAP as 50 kVp 1.6 mAs^{-1} and 60 kVp 0.8 mAs^{-1}, yet the CNR for bone is significantly superior at 40 kVp 4 mAs^{-1}.

Additional beam filtration

At a given kVp, additional beam filtration removes lower energy photons, thereby raising the average beam energy. At a constant kVp and detector exposure, additional beam filtration will: increase mAs, reduce entrance skin dose, reduce effective dose by a smaller amount (depending on projection) and reduce CNR (Martin 2007, Hess *et al* 2011, Smans 2009, Brosi *et al* 2011). Several studies have recommended differing thicknesses of added beam filtration for paediatrics and adults of up to 3 mm Cu (Smans 2009, Brosi *et al* 2011, Hamer *et al* 2005). The

Table 7.1. DAP (cGycm^2) at varying kVp and mAs at constant collimation (30 × 20 cm) and tube-to-image distance (100 cm).

Tube load (mAs)	40 kVp	50 kVp	60 kVp
0.8			0.88
1		0.60	1.10
1.6	0.17	0.96	1.76
3.2	0.73	1.92	3.53
4	0.94	2.40	
5	1.23	3.01	

European Guidelines recommended 0.1 mm Cu + 1 mm Al or 0.2 mm Cu + 1 mm Al additional beam filtration for use with film/screen and older generator technology (EUR16260 1996b, EUR16261 1996a, Neitzel 2004). In older generation technology, additional beam filtration is also required for some generators that are not able to cope with very short exposure times resulting from the high kVp technique. Additional beam filtration is selected automatically when using the appropriate projection and age group exposure preset. ICRP Publication 121 does not recommend use of additional beam filtration in neonates and very small infants due to the low kVp used (ICRP Publication 121 2013). This is inconsistent with the European Guidelines and other literature (EUR16261 1996a, Smans 2009). To improve CNR, further optimization of kVp, additional beam filtration and mAs may be required for body exposures as well as femur and humerus exposures in children up to 3 years old, whereas 0.1 mm Cu + 1 mm Al additional beam filtration appears to compromise CNR. For distal extremity exposures, additional beam filtration filters the lower energy photons resulting in lower contrast that affects CNR (Hess *et al* 2011, 2012). Thus, no additional beam filtration is recommended for distal extremity exposures.

Collimation and shielding

Appropriate collimation to the limits of the anatomy under investigation reduces radiation dose to paediatric patients. In limiting the exposed anatomy, the patient's dose is reduced and scatter radiation to the image receptor is minimized. Appropriate collimation is critical in DR because digital image detectors are more sensitive to low levels of radiation and the resulting images might show reduced contrast because of additional radiation scatter detected. Few studies have shown that poor collimation of lumbar spines has led to unnecessary radiation exposure for children. In DR, a dedicated software tool enables electronic masking based on the recognition of the exposed area borders. Adjustment of the electronic masking may be required so that the mask and exposed area are accurately aligned and the areas with a high brightness level on the image display monitor, that affect viewing conditions, are reduced. Masking should not be used as a replacement of physical beam collimation of the X-ray field size and should never be used to cover the anatomy that is contained within the exposed field. When the image receptor is to be exposed with multiple fields, the exposure fields need to be aligned, avoiding overlapping areas. Lead shielding on areas of the receptor not being exposed by the X-ray beam should be also used. Appropriate shielding also can help protect children's radiosensitive organs. In particular, specially formatted shields can reduce the radiation dose to male gonads and female breasts. Shielding is particularly important to protect anatomic areas near the exposure field, but should not affect the diagnostic value of the image. Patient's gonads should be shielded within 5 cm of the edge of a well collimated beam. This is particularly important in DR because shielding can affect the system's ability to optimize the display of the region of interest when the shielding material is included partially in the imaged area. Appropriate shielding is an important radiation safety practice in DR. Radiographers should follow department protocols regarding collimation and shielding for paediatric examinations.

Patient positioning

Correct positioning is important in radiographic/fluoroscopic image quality. Several studies have identified positioning errors as the major reason for having to repeat a radiography examination since the increased exposure latitude of digital image detectors has reduced repeated examinations compared to conventional screen-film radiography. Inaccurate positioning of the anatomy under investigation relative to the image detector, along with a poorly collimated X-ray beam, results in low quality radiographic images. Paediatric patients in particular, have more difficulty in complying during positioning and image capturing. Inaccurate positioning may affect how the system's software generates the image. Immobilization devices help ensure that paediatric patients can be imaged without retake. However, care needs to be taken since immobilization devices can cause image artifacts. Every repeated exam that is avoided represents a 100% patient radiation dose saving. Another leading cause of nondiagnostic images and repeated examinations is inadequate inspiration, especially in bedside exams (Blado *et al* 2003). Assuming that these two errors constitute about 50% of repeats, and that the total repeat rate might be about 5%, eliminating all these repeats would only reduce the total dose delivered to the patient population by 2.5%.

Detector exposure, AEC and mAs manipulation

A literature review reveals that there is limited information on detector exposures typically used in digital paediatric radiography in clinical practice. This may be due to the variations in technique and available equipment used at different departments. The IEC62494-1 standard EI is based on the average segmented detector exposure multiplied by 100 (IEC62494-1 2008). Thus, at a detector exposure of 2.5 μGy the EI should theoretically be 250. This is equivalent to a speed class of 400. The sensitivity of AEC systems can be tuned by altering the detector exposure or speed class at which the exposure is terminated. For paediatric body exposures, target detector exposure may be generally reduced from 2.5 μGy to 2 μGy (~500 speed), allowing for a 20% reduction in mAs and a simultaneous improvement in SNR. Repeat exposures should rarely be required unless the detector exposure is <1 μGy (~1000 speed).

The use of AEC detector cells is often problematic in paediatric radiography, particularly in cases when the body part being imaged is smaller than the area covered by the active AEC cells (Goske *et al* 2011). In some cases, AEC may be used in children only if the central AEC cell is enabled and the child's body anatomy is positioned so that that cell is entirely covered. It should be emphasized that in younger children the examined anatomy may be even smaller than the area of the active cell. Manual technique exposure charts may thus be more appropriate in younger children and in children with a small body habitus. To use manual technique charts dedicated for children, medical physicists need to develop them. To establish technique exposure charts for common radiographic projections, body weight or body part circumference is used to tailor the mAs and kVp for each patient (EC 2018). Patient age should be discouraged to be used as a guide to develop these charts. Typically, the abdomen of the largest 3-year-old child may be of the same size as the abdomen of the smallest 18-year-old child (Kleinman *et al* 2010). Radiography equipment vendors have different processing algorithms for children and adults for the same body region (Monnin *et al* 2005). The programmed technique exposure charts for children must be periodically reviewed and modified when necessary to assure that appropriate exposure parameters are selected for both AEC and manual technique selection. For body exposures, mA should be as high as possible at the small focal spot to maximize image sharpness and reduce exposure time. The large focus should generally be used for large body anatomies such as spine or obese adolescents. For distal extremity exposures, the kVp, mAs combinations should yield detector exposures in the range between 2.5 μGy and 5 μGy depending on projection orientation and clinical indication.

7.1.6 Diagnostic reference levels (DRLs) in paediatric radiography and fluoroscopy

DRLs are a useful tool in the attempt to optimize paediatric patient doses in radiography/fluoroscopy. Recently, the PiDRL workshop has indicated that only a few countries have set DRLs for paediatric examinations and there is lack of NDRLs for many examinations, in particular, for paediatric interventional procedures (EC 2018). Furthermore, the existing DRLs are often adopted from the old EC guidelines, while only a few countries have established DRLs based on their own national patient dose surveys. Due to the large variation of patient sizes within the paediatric population, several age and weight groups are needed to establish reliable DRLs. Tables 7.2–7.4 tabulate NDRLs values for paediatric head, chest and abdomen/pelvis examinations and procedures performed in various European countries.

Recently, the PiDRL project has given basic recommendations on how to establish and use DRLs for paediatric X-ray examinations. The following key points for establishing DRLs have been introduced:

- The quantity that should be used to establish DRL is P_{KA} and $K_{a,e}$ for radiography and P_{KA}, $K_{a,r}$, fluoroscopy time and number of images for fluoroscopic procedures.
- The parameter to group the patients should be patient weight for all body examinations and patient age for all head examinations. Grouping of patients should be carried out with *intervals* as follows:

  Weight groups for body exams: <5 kg, 5–<15 kg, 15–< 30 kg, 30–<50 kg, 50–<80 kg.

  Age groups for head exams: 0–<4 weeks, 5 weeks–< 1 y, 1–<6 y, ⩾6 y.
- NDRLs should be based on national patient dose surveys with a representative sample of all radiological institutions and all types of equipment and practices in the country when practical. From each hospital or radiology department a representative sample of at least 10 patients per procedure type and per patient group is recommended for radiography examinations, and at

Table 7.2. Paediatric national DRLs for head radiography in various European countries.

		Examination			
		Head, skull AP/PA		Head, skull LAT	
Country	Age group (y)	Entrance-surface air kerma, $K_{a,e}$ (mGy)	Kerma-area product, P_{KA} (mGy·cm^2)	Entrance-surface air kerma, $K_{a,e}$ (mGy)	Kerma-area product, P_{KA} (mGy·cm^2)
Austria (Billiger *et al* 2010)	0	0.35	150	0.30	100
	1	0.60	250	0.40	200
	5	0.75	350	0.50	250
	10	0.90	450	0.55	300
	15	1.00	500	0.60	350
Spain (Ruiz-Cruices 2016)	0		130		
	1–5		230		
	6–10		350		
	11–15		430		
Ireland (Hart *et al* 2000)	5	1.37		0.82	
Italy (EC Radiation Protection 109 1999)	5	1.5		1.0	
Poland (EC Radiation Protection 109 1999)	5	1.5		1.0	

Table 7.3. Paediatric national DRLs for thorax radiography in various European countries.

	Examination		
	Thorax PA		
Country	Age group (y)	Entrance-surface air kerma, $K_{a,e}$ (mGy)	Kerma-area product, P_{KA} (mGy·cm^2)
Austria (Billiger *et al* 2010)	0	0.05	
	1	0.06	
	5	0.07	
	10	0.09	
	15	0.11	
Spain (Ruiz-Cruices 2016)	0		40
	1–5		50
	6–10		85
	11–15		100
Ireland (Hart *et al* 2000)	1	0.057	
	5	0.053	
	10	0.066	
	15	0.088	
Italy (EC Radiation Protection 109 1999)	5	0.1	
Poland (EC Radiation Protection 109 1999)	5	0.1	

least 20 patients per procedure type and per patient group for fluoroscopy and fluoroscopically guided procedures.

- In collecting the patient dose data for the DRLs, likewise in daily imaging practices, there should always be a system in place to judge whether *image quality* is adequate for the diagnosis according to the indication of the examination.
- Besides the patient dose data according to patient grouping, *other data* such as X-ray equipment type, exposure parameters and use of AEC, should be collected for the evaluation and decision making.
- Patient dose surveys should be conducted by the authoritative body that sets the DRLs with the collaboration of authorized clinical experts.
- The comparison of local patient dose levels of a hospital or a group of hospitals with local DRLs or NDRLs should be carried out at the minimum frequency of once per year. A median value of the local patient dose distribution from a sample of at least 10 patients per patient group from each hospital should be used to compare against the DRL.

Table 7.4. Paediatric national DRLs for abdomen, pelvis and micturating cystourethrography (MCU) X-ray procedures in various European countries.

	Abdomen, common technique			Pelvis		MCU
Country	Age group (y)	$K_{a,e}$ (mGy)	P_{KA} (mGy·cm^2)	$K_{a,e}$ (mGy)	P_{KA} (mGy·cm^2)	P_{KA} (Gy·cm^2)
Austria (Billiger *et al* 2016)	0	0.20	60			0.5
	1	0.30	90			0.7
	5	0.40	200			1.2
	10	0.75	500			2.0
	15	1.00	700			
Spain (Ruiz-Cruices 2016)	0		150		60	0.5
	1–5		200		180	0.75
	6–10		225		310	0.90
	11–15		300		400	1.45
Ireland (Hart *et al* 2000)	1	0.33		0.26		0.9
	5	0.75		0.47		1.1
Italy (EC Radiation Protection 109 1999)	5	1.0		0.9		
Poland (EC Radiation Protection 109 1999)	0			0.2		
	5	1.0		0.9		

- NDRLs should be updated every 5–8 years for radiography and fluoroscopy. Local DRLs should be updated more frequently if there are changes in the equipment or practices that have a potential impact on patient dose levels.
- The NDRLs should be compared with available European DRLs whenever either of the values have been established or updated.

Table 7.5 tabulates European paediatric DRLs for Radiography and Fluoroscopy adapted from the European Guidelines on DRLs for Paediatric Imaging (EC 2018).

It should be noted that European DRL values tabulated in table 7.5 are indicative and should only be considered as a preliminary guideline at the national level. Instead of adopting these values, it is recommended that each country should establish NDRLs based on complete national patient dose surveys. NDRLs based on a country's national patient dose surveys should not exceed the values of the European DRLs. If this is so, specific action should be taken to optimize the radiographic/fluoroscopic procedure and establish a new NDRL that better conforms to the corresponding European DRL.

Table 7.5. European paediatric DRLs for radiography and fluoroscopy. Adapted from European Guidelines on DRLs for Paediatric Imaging (EC 2018).

Examination	Weight group (kg)	Age group (y)	Entrance-surface air kerma, $K_{a,e}$ (mGy)	Kerma-area product, P_{KA} ($mGy{\cdot}cm^2$)
Head	10–15	1		230
	15–30	5		300
Thorax PA	<10	0		14
	10–15	1		20
	15–30	5	0.08	39
	30–60	10	0.11	38
	>60	15	0.11	73
Thorax LAT	15–30	5	0.14	40
	30–60	10		60
Abdomen	10–15	1		150
	15–30	5	0.75	250
	30–60	10		425
Pelvis	15–30	5	0.48	
MCU[a]	<10	0		300
	10–15	1		700
	15–30	5		800
	30–60	10		750

[a] MCU: micturating cystourethrography.

7.1.7 Practices for optimization of fluoroscopic examinations performed on paediatric patients

In fluoroscopic examinations, the patient should be placed close to the image detector, with the tube as far from the patient as possible to minimize local entrance skin dose and reduce the effect of geometrical unsharpness. This is of paramount significance when using automatic brightness control. In interventional radiological (IR) procedures, this is particularly important as long as fluoroscopy times and multiple exposures can be anticipated. Automatic brightness control may need to be switched off during fluoroscopic examinations where there are relatively large areas of positive contrast material to avoid excessive dose rates. In the supplementary fluoroscopic examinations of the urinary tract, a grid is rarely necessary. Only fluoroscopic equipment with the potential for quick and easy removal of the grid should be used in paediatric age groups. Local skin dose can be high if the same projection is maintained throughout the whole or a large fraction of the procedure. Changing the projection can reduce the local skin dose below that for deterministic skin injuries but will not necessarily reduce the dose to internal organs or the stochastic radiation risk. To further reduce local skin dose, additional copper filtration can dynamically be inserted into the X-ray beam provided the generator is sufficient.

Pulsed fluoroscopy and electronic magnification
In the pulsed mode, the X-ray beam is switched on and off once during each fluoroscopic image; the pulse width, or duration, of each fluoroscopic image is shorter than the 33 milliseconds used in continuous fluoroscopy and typically ranges from 3 to 20 milliseconds to diminish motion unsharpness. However, the imaging unit must encompass technology that permits rapid starting and stopping of each pulse to avoid the tails ('ramp and trail effect'), which effectively increases the pulse width with an associated decrease in motion sharpness and elevation in patient dose.

In paediatric imaging, it is tempting to use electronic magnification to better visualize smaller structures. This can be achieved by decreasing the field of view, which increases sharpness of the small structures and enlarges them on the display. However, this is at a cost of increased radiation dose to the patient. On some older equipment, the patient dose doubles or even quadruples when the field of view is halved (Strauss 2006, Dixon *et al* 2006).

Fluoroscopic image hold and store
'Last image hold', is the display of the last fluoroscopic image on the display monitor when the fluoroscopy pedal or hand switch is released to terminate the fluoroscopic run. By use of this function, the radiologist can spend as much time as necessary studying the anatomy and other findings, without incurring additional radiation dose to the patient. Alternatively, 'fluoroscopy store' allows the operator to permanently store the fluoroscopic image on the reference monitor of the fluoroscope. Because the radiation dose to the image receptor for a fluoroscopic image is more than 10 times lower than the radiation dose used for a recorded digital image, the stored fluoroscopic image will contain more than three times the noise of the recorded digital image. If this level of image quality is tolerable, the additional recorded digital images and their comparatively high dose to the patient can be avoided especially in children, whose capacity for cooperation is frequently limited.

Collimation under fluoroscopic guidance
In paediatrics, collimation can be challenging because patient motion may place the area of interest outside the collimated field of view. Patient preparation and suitable immobilization are important to allow collimation to be effective. Alignment of the patient's anatomy of interest within the area of the X-ray beam should be initially completed without fluoroscopic guidance. Newer imaging equipment may provide an on-screen visualization of the collimator's blade's position superimposed on a 'fluoro-hold' image on the monitor, preventing the use of fluoroscopy for the collimation process. Finally, too much collimation may increase the patient's radiation dose. The fluoroscope measures the radiation dose reaching the image receptor by sampling typically 30%–50% of the area of the field of view. If the collimator blades enter this area, the correct dose rate to the image receptor appears to be too low because some of the detector areas do not receive radiation. This causes the unit to increase the kilovoltage and/or the tube current to compensate for this loss, which needlessly elevates the patient radiation burden.

This can be avoided on devices used for paediatric imaging by setting the sampling area to no more than 30% of the field of view.

7.1.8 Development of a quality assurance program

It is critical that radiologists, technologists and medical physicists develop quality standards for their institution aiming to assure diagnostic image quality at a properly managed radiation dose for paediatric patients. The literature has shown that up to 40% of the digital radiographs obtained from a single adult center are overexposed (Gibson *et al* 2012). This same center noted also that exposure creep was occurring in radiographs performed in intensive care units. Exposures in 43% of radiographs in a paediatric center using CR have been also reported to be overexposed (Don 2011). By recording and monitoring EIs, an individual hospital can control and reverse exposure creep (Gibson *et al* 2012). Analyzing the percentage of images that lies within and outside an acceptable range can be used to train technologists and thus decrease variation while improving image quality. With the advent of DR, much information is contained in the header of the digital imaging and communications in medicine (DICOM) image, which can be exported and used in a quality assurance program. Integrating the Healthcare Enterprise (IHE) radiation exposure monitoring (REM) profile also makes additional information available (O'Donnell 2011). This is a convenient method for routine evaluation of the performance metrics of DR for quality assurance analysis. Combined with the new IEC standard, it can be used by any hospital to develop its own quality assurance program. Cohen *et al* used the new IEC standard for monitoring EI in a neonatal intensive care unit over a 3-month period, and it showed no tendency for exposure creep (Cohen *et al* 2011).

7.2 Methods of dose optimization in CT

7.2.1 Introduction

CT scanning has such incredible value in diagnosis that it has changed radiology and several other specialties. Multi-detector-row technology, with sub second acquisition and CT fluoroscopy have boosted CT applications even more, enabling IR procedures that were traditionally performed with C-arm X-ray units. The continual increase in the number of slices during one rotation of the X-ray tube has brought multi-detector computed tomography (MDCT) into dynamic imaging and is now playing an important role in angiography. The development of hybrid systems such as PET/CT, SPECT/CT, CT simulators in radiotherapy and their incorporation in CT planning and dose delivery systems is moving CT from the domain of diagnostic radiology to other specialties. All the data show the steadily increasing impact of CT on medical exposures over the last decades, resulting in steadily increasing radiation doses. This trend seems to be attended by a decrease in the frequency of diagnostic radiology X-ray exams. Low dose imaging techniques such as conventional X-ray exams are steadily being replaced by high dose CT exams, resulting in an observed increase of medical exposures.

There is also growing realization that very often CT image quality is much higher than actually required to produce accurate clinical diagnosis. A number of studies reported large dose reductions using modified exposure parameters. However, the wide range in the exposure parameters reported, as well the different CT applications, reveal the difficulty in standardizing CT procedures. In the next few sections, specific practical ways of CT optimization methods are presented.

7.2.2 Justification as the first step of reducing radiation doses

Before any optimization is attempted, proper justification must be ensured. As the best optimization is no irradiation, justification should be the first principle to be applied when a request for CT examination is received by the radiologist. However, inadequate effort has been applied to justification and there is paucity of information on evaluation of the results of these efforts. The main reason for this is a culture of use of radiology for the purpose of defensive medicine. There are economic and political drivers that favor continuation of weak justification including:

- Target-driven processes;
- Clinical pathways;
- Self-referrals;
- Re-imbursement patterns;
- Financial models for the development of radiological services;
- Significant and systematic communication failures between healthcare professionals and both patients and the public (in common with other areas of medicine);
- A lack of clarity on the respective role of the radiologist versus the referring physician.

While most of the recommendations ask for joint decision making, there has been varied interpretation of the responsibility and the role. At some institutions, radiologists feel that it is their responsibility to accept, reject or advise on alternative techniques and at other hospitals, it is just the opposite. Everyday experience shows that clinicians tend to rely on imaging where clinical examination would have been regarded as sufficient. This is especially so for doctors in training who find an authoritative result reassuring in the face of inexperience in clinical skills. One should always keep in mind though that physicians are often subject to significant pressures (some country specific) from the medical system, the medico-legal system and from the public to prescribe CT. This is even more problematic in paediatric patients. A substantial number of paediatric CT scans lack appropriate justification. These can be replaced with other imaging modalities with lower or no ionizing radiation, such as radiography, ultrasonography or MRI. These exams can provide similar diagnostic information for some clinical indications. Ultrasonography, for example, can provide all the necessary diagnostic information for several clinical queries in paediatric clinical problems in the abdominal, neck, or musculoskeletal regions.

Appropriateness criteria (AC)

There is a lot of information in the American College of Radiology (ACR) Appropriateness Criteria (AC) document. These are evidence-based guidelines to assist referring physicians and other providers in making the most appropriate imaging or treatment decision for a specific clinical condition. Relevant information can also be found on the website of the Royal College of Radiologists.

The introduction of informed consent for patients undergoing CT scanning with regards to potential contrast media or radiation risks, may help in creating greater awareness among patients and greater responsibility for requesting physicians and radiologists. In a number of surveys conducted in the USA in recent years, controversial results are presented. It has the potential to increase informed, shared decision making for patients, as well as to reduce the risks and costs associated with the CT procedure. There are physicians that believe that informed consent for communicating the risk of radiation-induced cancer should be obtained from patients undergoing radiation based imaging. Furthermore, patients, physicians and radiologists alike were unable to have a sense of reasonable estimates of CT doses regardless of their experience level. However, a survey performed in 113 radiology departments in US academic medical centers showed that although most of them had guidelines for informed consent regarding CT, only a minority of institutions informed patients about possible radiation risks and alternative exams. Patients were not given information about the risks, benefits and radiation dose for a CT scan. The literature today has limited information on informed consent in CT. There are a number of components of informed consent, including information provision as well as social, personal and cultural factors that will we possibly need to address. There are also situations where informed consent is not possible (an unconscious patient or non-availability of next of kin). In such situations, the physician decides if the benefit outweighs the risk and performs the procedure. Having mentioned all these above, a balance has to be sought so that patients who are informed of the radiation risks do not decline a CT that may be essential to their diagnosis and subsequent treatment.

The Image Gently Choosing Wisely website (https://imagegently.org/) has a lot of information that helps referring physicians to choose tests and examinations that are evidence based, that do not duplicate other tests, are not harmful and are truly indicated. It is also very user-friendly for interested patients to learn more about the clinical issues. There are also systems that help the administration of hospitals to prevent unnecessary repeated CT and other radiological exams. They alert, for example, physicians about the number, type, and date of prior CT examinations performed at any imaging facility to prevent duplication in the case of CTs done recently. There are also software programs that permit CT studies that are done in different hospitals or institutions to be imported into the medical record of the patient, thus eliminating the possibility of doing the CT again.

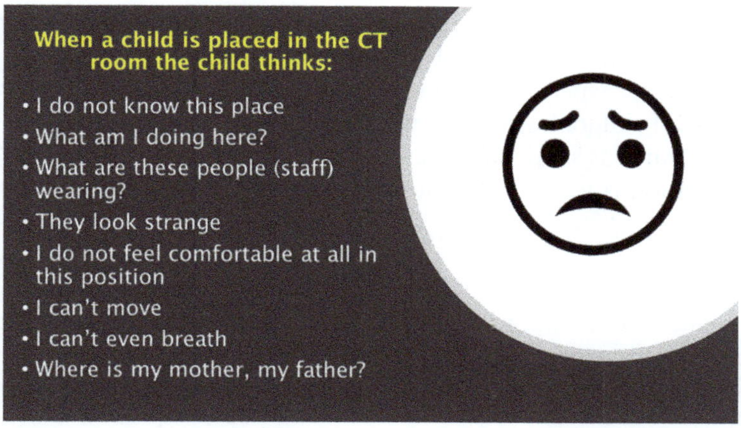

Figure 7.2. Stressed and/or frightened children usually have many questions when they enter the X-ray room for an examination.

7.2.3 Positioning of the paediatric patient

Comfortable and secure environment

The most important thing to consider when trying to position a child in the CT unit is that they will be stressed and afraid.

When a child is placed in the CT room they are scared (figure 7.2).

Additionally, parents could also be anxious. It is important that the environment is friendly, calm, with appropriate temperature and nice drawings, photographs or other material that children are familiar with. The staff should also be trained to provide precise instructions before scanning. There should be enough time; no hurry. In the case of available resources, a mock unit could be available for pre-scan adaptation. During CT scanning, distraction techniques or devices to improve cooperation could be used to reduce the child's anxiety and thus motion that could reduce the quality of the image; for example, projectors with child-friendly images, toys with flashing lights or music, child-friendly images on the ceiling and/or walls, a parent reading a favorite story or talking to them through the console comfort the child.

Exercising respiration commands

Exercising respiration commands before scanning could be helpful. If breath holding is not possible, then age-adapted instructions for quiet and superficial breathing should be given. Soft restraining techniques could also help especially in small children. A simple trick for saving radiation in a chest CT is to ask the patient to have his/her arms up.

Should one consider sedation?

Despite all efforts, sedation or anesthesia are sometimes unavoidable. It has to be considered as movement can decrease the diagnostic output of the CT scan. In the case of really life-threatening conditions such as paediatric stroke, where children are usually dizzy and not cooperative, CT and CT brain perfusion are not possible without sedation. In the case of excessive movement, procedures are often rescheduled until an expert sedation service provider is available. There are studies in the literature of children who underwent sedation or were given general anesthesia in which sedation was inadequate in 16% of children and failed in 7% of cases. Therefore, sedation is very important for successful CT scanning.

Patient centering
The *z* axis of the patient should be in the middle of the CT gantry as this will effectively reduce the radiation dose to the child (entrance skin dose to the patient is a function of the distance of the X-ray tube from the patient's skin).

7.2.4 Scanogram

The first radiation dose-influencing factor is the scanogram, also called a scout view or topogram. For optimization purposes there are various factors to be considered; three of which are the most important.

Tube position (above or below the table)
About 2/3 of radiation dose can be saved by just placing the tube below the CT couch. This may require reprogramming of the scanogram view default by the manufacturer and must be discussed with engineers of the purchased CT machine. This is very important as it reduces organ doses to radiosensitive organs of the patient such as male gonads, breast, thyroid and the lens of the eye.

Scanogram length
It should be as short as possible, tightly adapted to the region of interest.

Technical parameters
The kV and mAs settings can account for more than 50% of the dose in a single chest CT examination for paediatric patients as reported by Perisinakis *et al* (2001) for adults patients. Technical parameters should be tailored to the paediatric patient being scanned. Typical values proposed by Singh (2015) are 10–20 mAs and 80 kVp.

7.2.5 Shielding of organs for paediatric patients

Shielding could be used for organs such as the thyroid, lens of the eye and breast when they are not in the primary beam. Dose reductions reported are in the order of 40%–80%, or even 95% when shielding the testes in abdominal procedures. Breast-anlage (primordium or the first rudiment of the breast, the underdeveloped tissue) protection using 2 mm thick bismuth coated latex shielding reduces the dose to the breast-anlage by approximately 40%. One must keep in mind that these shields are meant for the external radiation scatter, whereas the internal scatter can hardly be affected. Therefore, despite the fact that testicular capsules can be appropriate in shielding, the deep location of the ovaries excludes any local protection.

Bismuth protection should only be placed after the scout view or scanogram (or AEC pre-scanning) is performed so that the system does not inappropriately increase tube current in the area of the shield. Lead-equivalent eye glasses in the case of high eye lens dose CT (such as CT of the brain and facial bones when angulation of the gantry is not enough to keep the orbits outside the examination volume) are also an alternative to bismuth shields but must be used with caution as their radiation protection rating has not been standardized internationally to date. If the patient is cooperative, the absorbed dose can be reduced by 50%–70%. One should always remember that streak artifacts and increased noise may result from sub optimally placed shielding.

7.2.6 Immobilization

Immobilization is required in many infants and young children when performing radiographic studies. Devices, such as foam rubber devices, may be used in very small infants. It may be useful to take advantage of the period when an infant is calm or asleep after a feed to perform the examination. Immobilization devices should be easy to use, and their application should not be traumatic to the patient (or caregivers). Their use and benefits should be explained to the accompanying caregiver. When physical restraint by parents or other accompanying persons is unavoidable, they should be informed about the exact procedure and what is required of them. They should be provided with a protective apron.

7.2.7 Paediatric specific exposure parameters

CT technology

The evolving technology continuously provides operators with software or hardware tools for the best CT dose optimization. Dual source CT scanning, for example, allows very quick scanning, minimizing exam times that could result in unacceptable image quality due to motion artifacts or noise. Another recent development is also the 320-row detector technology that facilitates volume CT and helps one to avoid the over-beaming effect of helical scanning; see Sorantin (2013b). This particular technology significantly contributes to dose minimization for paediatric CT, speeds up investigation and thus decreases sedation needs and duration.

As reported by Sorantin (2013b), simplified dose adjustments or modifications are a bit tricky since children develop rapidly and CT systems differ considerably. Thus, only some rules of thumb can be applied. Before actually attempting paediatric CT radiation dose optimization, the reader can also consult the IAEA TECDOC-16212009 (IAEA 2009) that provides guidelines for image quality assessment.

Clinical indication

Depending on the clinical indication the operator should select the most appropriate CT anatomical program to scan the patient. Before scanning, the operator should verify that the technical parameters of this particular program are appropriate for the particular indication and patient size. Patient size can be accurately determined by the measurement of the lateral dimension of the patient laying supine on the CT gantry couch as mentioned in the AAPM Report 96 (AAPM 2008). For the purpose of minimizing radiation exposure, noisier images, if sufficient for radiological diagnosis, should be accepted. According to Triantopoulou (2017), paediatric patients with malignancies are those exposed to higher levels of radiation during CT imaging and more studies are needed for the determination of radiation dose in those patients. According to the study, the main pathologies for which CT dose is reported are: Crohn's disease, hydrocephalus, cystic fibrosis and paediatric malignancies—mainly lymphoma. The related radiation dose data are extremely scarce and variable (0.2–518 mSv).

Tube potential
Almost all manufacturers nowadays provide multiple choices of tube potential (kVp). Until a decade ago, CT machines with fixed kVp in the range of 110–130 were standard practice. In recent years there has been greater realization that kVp can play a role in patient dose management in CT in the same way as in radiography but in the opposite direction. Higher kVp in radiography is recommended for dose reduction but lowering kVp is recommended for dose reduction in CT. The kVp in CT is already in a much higher range (100–140) than radiography (50–95). A reduction from 120 kV to 90 kV can result in as much as a 35% reduction in the radiation dose, without sacrifice of low-contrast detectability. Kalender (2009) also concluded recently that voltage settings in CT should be varied more often than is common in practice today and should be chosen not only according to patient size but also according to the substance imaged in order to minimize dose while not compromising image quality. As a general rule, 80 kV could be used for newborns and 100 kV later on until puberty. A good starting point for daily practise using 80 kV is to raise the mAs setting to about 175% of the 120 kV settings and then lower the mAs step by step. Due to the inherent higher contrast of 80 kV compared to 120 kV (mainly for contrast media application and for bone diagnostic) this would be acceptable (Sorantin 2013a). The reduction of kVp will also result in improved contrast on images, especially when an iodinated contrast medium is used. The increase in noise due to decreased photon penetration would not be of concern due to the paediatric patient's smaller size. Furthermore, due to the use of special algorithms such as the iterative algorithms that will be described below, these effects can be suppressed or eliminated. Once it is decided to lower the kVp, the optimization team should take into account also the effect of lower tube voltage on patient skin dose, the higher possibility of metallic artifacts, the decrease in detectability of low-contrast tissues, etc.

Tube current and tube current modulation
Tube current modulation is an extremely useful tool found in the current generation of CT scanners. It automatically adjusts mAs during a single rotation or part of the rotation of the CT tube around the patient according to the patient's anatomy in each direction of the tube. In simple terms, it means that during the posterior-anterior projection, less mAs can be used, whereas in lateral projections (the

attenuation from the patient being higher), the mAs increases accordingly. The specific tube current modulation techniques vary according to the manufacturer, but all major vendors provide this technology. It is reported that tube current modulation can reduce the radiation dose up to 40% per examination and can be used in most CT protocols. New organ-based dose modulation proposed by many manufacturers can reduce the mA over an arc of 120° anteriorly when the patient is supine for dose savings in the breast, thyroid, or lens. It should be noted, however, that tube current modulation does not totally free the operator from a selection of scan parameters and awareness of individual systems is important. While CT systems without this feature require operator selection of mA or mAs, tube current modulation systems require an understanding of concepts such as noise index, reference mAs and reference images in order for the feature to be operated effectively. Inappropriate patient centering can cause incorrect operation of the tube current modulation system, in which the tube current is controlled with information from localizer radiographs, and thus causes increases in tube current or image noise. Further investigation should be done to evaluate possible dose optimization when applying both tube current modulation and reduced kVp. An important thing to remember is that the shortest rotation time is generally required in paediatric CT as this will minimize motion artifacts.

In older CT scanners, where AEC does not exist, manual setting of mAs is required. The operator should be provided with appropriate guidelines for mAs selection in the abdomen CT as a function of patient size often referred to as a technique chart. If the technique chart is not available, then the mAs setting can be calculated using the weight (W) of the patient by the following simple formula (Baert 2007):

$$Q = \text{Tube current} \ast \text{rotation time} = (1.5\ \text{mAs kg}^{-1}) \ast W \qquad (7.1)$$

Technique charts for single detector and MDCT scanners of specific manufacturers are given in the American Association of Physicists in Medicine (AAPM) Report 96 (AAPM 2008). In this report, a relative technique chart for abdominal and pelvis CT in children is also given (table 7.6). This applies to any CT system or kVp, since all mAs values are normalized to the CT department scan protocol (shown in bold,

Table 7.6. Relative technique chart for abdominal and pelvic CT in children, for any CT system or kVp, since all mAs values are normalized to the CT department scan protocol (shown in bold, relative mAs = 1). The lateral patient width at the level of the liver is measured from the CT scanogram (AAPM 2008).

Lateral patient width (cm) (at the level of the liver)	mAs (relative to the protocol)
Up to 14	0.55
14.1–18	0.75
18.1–22	1.0

relative mAs = 1). The lateral patient width at the level of the liver is measured as mentioned above and in the AAPM Report 96 (AAPM 2008).

Slice collimation, pitch

The small size of a child requires relatively thinner slices compared with adults in order to improve spatial resolution. The slice collimation, especially in new generation MDCT scanners, should be as low as possible taking into consideration scan time and tube power. The problem is, however, that using identical exposure with thinner slices compared with thicker slices will automatically increase noise. To diminish noise and artifacts, it is recommended (a) to view thicker slices than collimation, (b) to use special reconstruction algorithms that diminish noise or (c) to perform multiplanar reformations (MPR) that erase most of the image noise. A similar approach can be applied for pitch selection, meaning that choice of pitch should be done with respect to scan speed, artifacts and power. For older single CT scanners, a pitch factor of 1.5 is recommended to reduce radiation dose. For new MDCT systems, the choice of pitch is a complex issue related to the CT technology and should be evaluated separately for each type of scanner. For example, in certain CT scanners, increases in pitch result in compensatory increases in tube current in order to keep the image quality at an acceptable level (and thus radiation dose). The operator must be aware of how the system operates in order to choose pitch accordingly.

Helical versus axial scanning

Head imaging has typically been performed in the axial mode. Axial scanning with the gantry tilted or proper positioning of the head during head scans could reduce radiation dose to the lens of the eye. Body imaging is performed in the helical mode in everyday clinical routine drastically reducing the scan time. In paediatric imaging, the advantages and disadvantages of axial and helical imaging must be carefully considered by the radiographer, radiologist and medical physicist; a helically-acquired head exam or axially-acquired body exam could be also chosen depending on the clinical case (IAEA 2014).

7.2.8 Filtration

X-ray filters are used in radiology for cutting off the X-rays that have lower energy and do not contribute to the image but only to the patient dose. Bone filters have the best spatial resolution, whereas soft tissue filters have lower spatial resolution. For clinical applications in which high spatial resolution is less important than high contrast resolution—for example, in scanning for metastatic disease in the liver, soft tissues can be applied. Bow-tie or beam shaping filters reduce radiation dose by 50% compared with conventional flat filters. Software noise reduction filters are also an alternative, especially in high contrast examinations such as chest CT. Bone kernels accentuate higher frequencies in the image at the expense of increased noise. Soft tissue kernels produce images with reduced noise but lower spatial resolution.

7.2.9 Iterative reconstruction algorithms

To optimize image quality at low dose settings, CT manufacturers have introduced new reconstruction algorithms. Iterative reconstruction methods acquire CT data at a much lower tube current and process raw data to lower image noise by performing multiple iterations with the goal of preserving image quality. The algorithms can reduce radiation dose by 40%–80% but the process is time consuming and readers need a period of adjustment to adapt to the image appearance. As technology evolves and computing power increases, the reconstruction of CT data into various image datasets seems to be no longer a major issue. As reported by Mayo-Smith (2014), hybrid iterative reconstruction methods that blend filtered back projection and iterative reconstruction are introduced into the market. These hybrid software perform the majority of image noise reduction in image space rather than raw or sinogram data resulting in an increase of image reconstruction speed so the clinical workflow is unaffected.

7.2.10 Limitation of scan length

The scanning length for a particular type of CT examination can vary due to:

1. Pathology;
2. Size of the patient;
3. Experience of user;
4. Demographics of a country (height of the population).

With the evolution of CT scanners, the introduction of MDCT and the dramatic reduction of rotation times to sub seconds, users are tempted to extend the region of interest beyond the one required. CT protocols need to be established so as to limit irradiation only to the particular body region under investigation. Only radiologists properly trained in CT as well as radiation protection issues related to the CT technique should perform such procedures. The large range in DLP values in the literature reveals the differences in technique in CT departments. Some operators examine the upper abdomen in cases of hepatic and pancreatic disease, whereas others examine the whole abdomen, which includes also the pelvic region. Many clinical studies must be performed so as to gain consensus for the optimal length of examination.

As a rule, the operator should always limit scan coverage. This applies for both the scanogram as well as the actual CT exam. He must also avoid non-justified multiple scans of the same area. If repeat scans are absolutely necessary, consideration should be given to limiting these to the smallest volume possible or performing them at a lower dose that will not obscure the additional information expected. Multiphase CT examinations in children should be justified in each clinical case.

7.2.11 Scan series

Most paediatric examinations require only one scan. Pre and post-contrast scans or delayed imaging rarely provide additional information and should not be performed

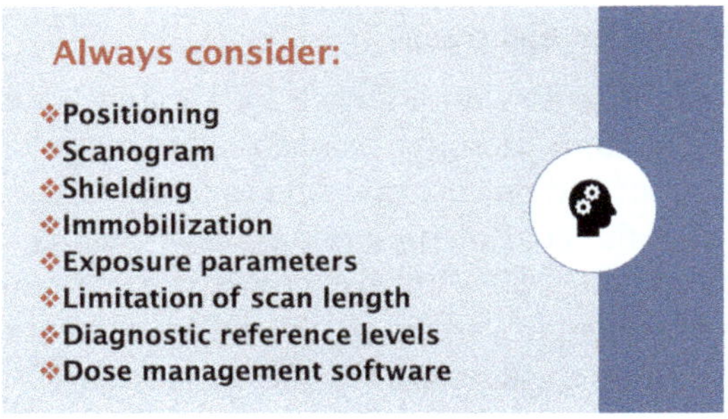

Figure 7.3. General rules for paediatric radiological optimization.

unless specifically indicated through consultation between the paediatric radiologist and the referring physician.

7.2.12 Diagnostic reference levels (DRLs) in paediatric CT

To assist in the optimization process, the concept of DRLs has been introduced. A DRL value is advisory, and in practice is set so that if the value is exceeded regularly, the practice involved should be investigated. This does not mean that there is necessarily unacceptable practice; rather, the practice requires explanation, review, or possibly a new approach. The most up to date information and guidance on how to set up CT DRLs can be found in the European Commission Report 185 (EC 2018) 'European Guidelines on Diagnostic Reference Levels for Paediatric Imaging' and in the recent ICRP Publication 135 (ICRP 2017) Report on 'Diagnostic Reference Levels in Medical Imaging'. More information about paediatric DRLs as a tool for optimization of radiographic and fluoroscopic examinations can be found in section 7.1.5.

7.2.13 Dose management software

Dose management systems are now available and facilitate the establishment of databases as repositories of dosimetric data, aid in radiation protection quality assurance and quality improvement. As these systems become more and more established, they allow large dose registries to be built up, they are helpful in fulfilling legal requirements such as European Union requirements for reporting dose results to authorities for clinical audits, or for following the European Union's basic safety standards directive to identify unintended overexposures. A number of publications have concluded already that these software provide a quick and effective tool for tracking and optimizing doses in radiological and CT examinations by allowing fast reviewing of exposures giving the opportunity to act quickly. The most recent robust examples are the publications of (1) Little (2015); the authors succeeded in reducing CT radiation doses by a maximum of 30% through a program that included protocol changes, iterative reconstruction, optimization of scan acquisition, technologist education, and continuous monitoring with feedback tools and (2) Niiniviita *et al* (2018); the main conclusion was that dose monitoring software offered a valuable tool for evaluating the imaging practices and finding non-optimized protocols in paediatric and young adults undergoing head and cervical spine CT. A summary of simple general rules for paediatric radiation dose optimization are found in figure 7.3.

References

ACR-AAPM-SIIM 2013 Practice guidelines for digital radiography *J. Digit. Imaging* **26** 26–37

American Association of Physicists in Medicine (AAPM) 2008 Task Group 23 Report *CT Dosimetry: The Measurement, Reporting, and Management of Radiation Dose in CT* AAPM Report 96, AAPM

Baert A L, Knauth M and Sartor K 2007 *Radiation Dose from Adult and Pediatric Multidetector Computed Tomography* (Berlin Heidelberg New York: Springer)

Billinger J, Nowotny R and Homolka P 2010 Diagnostic reference levels in pediatric radiology in Austria *Eur. Radiol.* **20** 1572–9

Blado M E, Ma Y and Corwin R A *et al* 2003 Impact of repeat/reject analysis in PACS *J. Digit. Imaging* **16** 22–6

Brosi P, Stuessi A, Verdun F R, Vock P and Wolf R 2011 Copper filtration in pediatric digital X-ray imaging: its impact on image quality and dose *Radiol. Phys. Technol.* **4** 148–55

Butler M L, Rainford L, Last J and Brennan P C 2010 Are exposure index values consistent in clinical practice? A multi-manufacturer investigation *Radiat. Prot. Dosim.* **139** 371–4

Carestream Health 2012 *Maximizing Dose Efficiency for Pediatric Patient Imaging* (Rochester, NYC: Carestream Health)

Chapple C L, Faulkner K and Hunter E W 1994 Energy imparted to neonates during X-ray examinations in a special care baby unit *Br. J. Radiol.* **67** 366–70

Cohen M D, Cooper M L, Piersall K and Apgar B K 2011 Quality assurance: using the exposure index and the deviation index to monitor radiation exposure for portable chest radiographs in neonates *Pediatr. Radiol.* **41** 592–601

Dixon R 2006 Special procedures (angiography): clinical practice *Categorical Course Syllabus in Diagnostic Radiology Physics: From Invisible to Visible—The Science and Practice of x-Ray Imaging and Radiation Dose Optimization* ed D P Frush and H Huda (Oak Brook, IL: Radiological Society of North America), pp 203–9

Don S 2011 Pediatric digital radiography summit overview: state of confusion *Pediatr. Radiol.* **41** 567–72

Don S, Whiting B R, Rutz L J and Apgar B K 2012 New exposure indicators for digital radiography simplified for radiologists and technologists *Am. J. Roentgenol.* **199** 1337–41

Don S 2004 Radiosensitivity of children: potential for overexposure in CR and DR and magnitude of doses in ordinary radiographic examinations *Pediatr. Radiol.* **34** S167–72

EC Report 185 2018 European Guidelines on Diagnostic Reference Levels for Paediatric Imaging, EU publication

European Commission 1996a *European Guidelines on Quality Criteria for Diagnostic Radiographic Images in Paediatrics. EUR16261* (Luxembourg: Office for Official Publications of the European Communities)

European Commission 1996b *European Guidelines on Quality Criteria for Diagnostic Radiographic Images. EUR 16260* (Luxembourg: Office for Official Publications of the European Communities)

Gibson D J and Davidson R A 2012 Exposure creep in computed radiography: A longitudinal study *Acad. Radiol.* **19** 458–62

Goske M J, Charkot E and Herrmann T *et al* 2011 Image gently: challenges for radiologic technologists when performing digital radiography in children *Pediatr. Radiol.* **41** 611–19

Hadley J L, Agola J and Wong P 2006 Potential impact of the appropriateness criteria on CT for trauma American College of Radiology *Am. J. Roentgenol.* **186** 937–42

Hamer O W, Sirlin C B, Strotzer M, Borisch I, Zorger N and Feuerbach S *et al* 2005 Chest radiography with a flat-panel detector: image quality with dose reduction after copper filtration *Radiology* **237** 691–700

Hart D, Wall B F, Shrimpton P C and Dance D 2000 The establishment of reference doses in paediatric radiology as a function of patient size *Radiat. Prot. Dosim.* **90** 235–8

Herrmann T L, Fauber T L, Gill J, Hoffman C, Orth D K, Peterson P A, Prouty R R, Woodward A P and Odle T G 2012 Best practices in digital radiography *Radiol. Technol.* **84** 83–9

Hess R and Neitzel U 2011 *Optimizing Image Quality and Dose in Digital Radiography of Pediatric Extremities* (Eindhoven: Philips Healthcare)

Hess R and Neitzel U 2012 Optimizing image quality and dose for digital radiography of distal paediatric extremities using the contrast-to-noise ratio *Fortschr. Rontgenstr.* **184** 643–9

IAEA 2009 IAEA-TECDOC-1621, Dose reduction in CT while maintaining diagnostic confidence: a feasibility/demonstration study, IAEA

ICRP 2007 The 2007 recommendations of the International Commission on Radiological Protection. ICRP Publication 103 *Ann. ICRP* **37** 2–4

ICRP 2013 Publication 121 Radiological protection in paediatric diagnostic and interventional radiology *Ann. ICRP* **42** 1–63

ICRP 2017 Publication 135 *Diagnostic Reference Levels in Medical Imaging Ann. ICRP* **46** 1–144

ICRU 1996 Report 54 *International Commission on Radiation Units and Measurement. Medical Imaging: The Assessment of Image Quality* (Bethesda, MD)

IEC 2008 Medical Electrical Equipment-Exposure Index of Digital X-Ray Imaging Systems-Part 1: Definitions and Requirements for General Radiography (Geneva, Switzerland)

International Atomic Energy Agency (IAEA) 2014 *Dosimetry in Diagnostic Radiology for Paediatric Patients IAEA Human Health Series* 24 (Vienna: IAEA)

International Atomic Energy Agency 2009 *Dose Reduction in CT While Maintaining Diagnostic Confidence: A Feasibility/Demonstration Study* Document no. IAEA-TECDOC-1621 (Vienna, Austria)

Kalender W A, Deak P and Kellermeier M 2009 Application- and patient size-dependent optimization of x-ray spectra for CT *Med. Phys.* **36** 993–1007

Kallman H E, Halsius E, Folkesson M, Larsson Y, Stenstrom M and Bath M 2011 Automated detection of changes in patient exposure in digital projection radiography using exposure index from DICOM header metadata *Acta Oncol.* **50** 960–5

Kleinman P L, Strauss K J, Zurakowski D, Buckley K S and Taylor G A 2010 Patient size measured on CT images as a function of age at a tertiary care children's hospital *Am. J. Roentgenol.* **194** 1611–19

Little B P, Duong P A, Knighton J, Baugnon K, Campbell-Brown E, Kitajima H D, Louis S St, Tannir H and Applegate K E 2015 A comprehensive CT dose reduction program using the ACR dose index registry *J. Am. Coll. Radiol.* **12 Pt A** 1257–65

Martin C J, Sharp P F and Sutton D G 1999 Measurement of image quality in diagnostic radiology *Appl. Radiat. Isot.* **50** 21–38

Martin C J 2007 The importance of radiation quality for optimization in radiology *Biomed. Imaging Interv. J.* **3** e38

Mayo-Smith W W *et al* 2014 How I do it: managing radiation dose in CT *Radiology* **273** 657–72
Monnin P, Gutierrez D, Bulling S, Lepori D, Valley J-F and Verdun F R 2005 Performance comparison of an active matrix flat panel imager, computed radiography system, and a screen-film system at four standard radiation qualities *Med. Phys.* **32** 343–50
NCRP 1981 Report No. 68 *National Council on Radiation Protection and Measurements. Radiation Protection in Pediatric Radiology* (Bethesda, MD)
Neitzel U 2004 Pediatric radiation dose management in digital radiography *Pediatr Radiol* **34** 227–33
Niiniviita H *et al* 2018 Dose monitoring in pediatric and young adult head and cervical spine CT studies at two emergency duty departments *Emerg. Radiol.* **25** 153–9
O'Donnell K 2011 Radiation exposure monitoring: a new IHE profile *Pediatr. Radiol.* **41** 588–91
Oikarinen H, Merilainen S, Paakko E, Karttunen A, Nieminen M T and Tervonen O 2009 Unjustified CT examinations in young patients *Eur. Radiol.* **19** 1161–5
Perisinakis K, Damilakis J, Voloudaki A, Papadakis A and Gourtsoyiannis N 2001 Patient dose reduction in CT examinations by optimising scanogram acquisition *Radiat. Prot. Dosim.* **93** 173–8
Ruiz-Cruces R *et al* 2016 Diagnostic reference levels and complexity indices in interventional radiology: a national programme *Eur. Radiol.* **26** 4268–76
Schaefer-Prokop C, Neitzel U, Vanema H W, Uffmann M and Prokop M 2008 Digital chest radiography: an update on modern technology, dose containment and control of image quality *Eur. Radiol.* **18** 1818–30
Seibert J A and Morin R L 2011 The standardized exposure index for digital radiography: an opportunity for optimization of radiation dose to the pediatric population *Pediatr. Radiol.* **41** 573–81
Seibert J A 2004 Trade offs between image quality and dose *Pediatr Radiol* **34** S183–95
Shepard S W 2009 An exposure indicator for digital radiography *Report of AAPM Task group* **116** 1–92
Singh S, Kalra M K, Thrall J H and Mahesh M 2012 Pointers for optimizing radiation dose in pediatric CT protocols *J. Am. Coll. Radiol.* **9** 77–9
Smans K 2009 *The Development of Dose Optimisation Strategies for X-Ray Examinations of Newborns* (Belgium: Katholieke Universiteit Leuven)
Sorantin E *et al* 2013a CT in children–dose protection and general considerations when planning a CT in a child *Eur. J. Radiol.* **82** 1043–9
Sorantin E *et al* 2013b Experience with volumetric (320 rows) pediatric CT *Eur. J. Radiol.* **82** 1091–7
Strauss K J 2006 Pediatric interventional radiography equipment: safety considerations *Pediatr. Radiol.* **36** 126–35
Triantopoulou S and Tsapaki V 2017 Does clinical indication play a role in CT radiation dose in pediatric patients? *Phys. Med.* **41** 53–7
Uffmann M, Neitzel U, Prokop M, Kabalan N and Weber M *et al* 2005 Flat-panel-detector chest radiography: effect of tube voltage on image quality *Radiology* **235** 642–50
UNSCEAR 2006 *Report. Effects of ionizing radiation. Volume I: Report to the General Assembly, Scientific Annexes A and B* United Nations. Scientific Committee on the Effects of Atomic Radiation, United Nations sales publication E.09.IX.5 (New York: United Nations)
UNSCEAR 2013a *Report to the General Assembly. Volume II: Scientific Annex B: Effects of radiation exposure of children*

UNSCEAR 2013b *Report. Sources, effects and risks of ionizing radiation. Volume II: Scientific Annex B: Effects of radiation exposure of children* United Nations Scientific Committee on the Effects of Atomic Radiation, United Nations Sales No. E.14.IX.2 (New York: United Nations)

US Department of Health and Human Services 2003 *Food and Drug Administration. Guidance for Industry and FDA Staff: Pediatric Expertise for Advisory Panels* (Rockville, MD: US Department of Health and Human Services, Food and Drug Administration, Center for Devices and Radiological Health)

World Health Organization (WHO) Communicating radiation risks in paediatric imaging. Information to support healthcare discussions about benefit and risk

Chapter 8

The management of (a) pregnant patients and (b) pregnant employees

John Damilakis

This chapter discusses what dose management entails and how basic management programs are implemented for (a) pregnant patients and (b) pregnant employees.

8.1 The management of pregnant patients

8.1.1 Introduction

Dose management can be defined as the organization and coordination of activities to justify an X-ray examination, estimate radiation doses and radiation-induced risks, communicate information related to doses and risks and optimize imaging acquisition protocols to ensure safety of the pregnant woman exposed to radiation.

Occasionally, X-ray diagnostic or interventional examinations are needed during pregnancy to provide information for significant medical problems and emergency situations. When this occurs, it is important to follow the right steps to minimize conceptus radiation dose. Pregnant patients are also exposed to X-rays during the first post-conception weeks before the diagnosis of pregnancy. These patients exposed accidentally need different dose management in comparison with those exposed intentionally. Furthermore, proper dose management is needed for pregnant staff working with X-rays.

8.1.2 Intentional exposure of pregnant patients

Essential elements of a program developed to manage a pregnant patient who requires radiological examination are (a) justification of the examination, (b) dose anticipation and estimation of radiogenic risk, (c) communication between the medical physicist and the referring physician and/or the radiologist, (d) communication with the patient and her relatives and (e) dose optimization.

doi:10.1088/978-0-7503-1317-9ch8

Justification of the exposure: benefit versus risk

X-ray examinations should be justified. The International Commission on Radiological Protection (ICRP 2000) states that *'After a type of examination or therapy has been justified generally, each specific instance should be justified. As an example, a standard radiotherapy protocol may be justified in a* 50*-year-old female but the same protocol may not be justified in a pregnant* 25*-year-old without more consideration and perhaps modification'* (ICRP 2000). The justification process includes a risk–benefit analysis. Benefit/risk assessment occurs at two levels. First, research teams evaluate benefits and risks for the population and provide information in the literature. This is the general justification process of a type of examination. Several articles provide conceptus dose and risk levels for radiation-induced effects resulting from X-ray examinations performed on the mother for the three periods of gestation. Information published in the literature is useful for referring physicians and radiologists to evaluate the benefits and risks of a specific examination for a patient. This is the second level of the benefit/risk assessment. In the case of pregnancy, expected benefits and potential risks of both the mother and the growing unborn child should be considered to select the examination of choice. Information about the benefits and risks should be properly communicated to the pregnant patient. A collaborative approach between referring physicians and radiologists is always useful to help explain radiogenic risks to pregnant patients who may have concerns about harmful effects of radiation.

It is always preferable to perform an examination based on non-ionizing radiation on pregnant women. In many cases, ultrasound or MRI can provide detailed information without exposing the patient to ionizing radiation. However, there are a number of situations in which there are indications that the use of X-rays is appropriate. In those cases, risk versus benefits analysis is needed.

In the case of pregnancy, there are two 'golden' rules that should always be taken into account:

1. When a pregnant patient requires an emergency radiologic examination, there should be no hesitation to do the study (Goldman and Wagner 1996);
2. While the risk of exposing the conceptus to X-rays is a primary consideration, the risk of delaying or not performing the radiologic examination must also be taken into account (Wagner *et al* 1997).

Conceptus risk is a function of gestational age and conceptus dose. Gestational age is the age of the pregnancy from the last normal menstrual period and can be determined accurately by using ultrasound. Radiation risks are related to the stage of pregnancy. These risks are most significant during the first trimester, less in the second trimester and least in the third trimester. During the first 2 weeks of conception, few damaged cells can be replaced by others and the conceptus will recover completely. However, if the radiation dose is high (higher than at least 50 mGy according to studies), many cells are damaged or killed from the exposure and the dose will terminate pregnancy. This is known as the 'all-or-nothing' period. In that situation, dose estimation is not needed because there is little to be gained from a dose evaluation. Also, if the conceptus dose is very low, detailed dose assessment is not needed. Extra abdominal diagnostic X-ray examinations are associated with very low conceptus doses. In general, conceptus dose from these studies is less than 1 mGy. Normally, it would be unnecessary to perform in-depth conceptus dose evaluation for such studies because the dose is negligible and there would not be any reason to alter patient management. Chapter 4 provides information on conceptus dose and radiation-induced risk associated with diagnostic and interventional X-ray examinations.

In the period from the 3rd to the 8th week there is a potential for malformation of organs. From the 8th to 15th weeks post-conception the most likely form of damage is mental retardation. The threshold doses for deterministic effects may be in the range of 100–200 mSv for acute exposure to the whole body. Therefore, these effects are unlikely to be observed after radio-diagnostic or even interventional studies. Stochastic effects are the only type of effect likely to be seen after these examinations. The magnitude of the risk of childhood fatal cancer following in utero exposure is 6% (0.06) Sv^{-1}. The magnitude of the risk of hereditary effects following in utero exposure is 1×10^{-2} Sv^{-1}. Of note, however, is that the risks per unit dose are derived from epidemiologic studies on effects of individuals exposed to high levels of ionizing radiation such as atomic bomb survivors. Therefore, the risk coefficients used for risk estimation have a wide range of uncertainty.

According to ICRP (ICRP 2000), installation-specific measurements and calculations of conceptus doses may be necessary if these doses are suspected of exceeding 10 mGy. Conceptus doses below 100 mGy should not be considered a reason for terminating a pregnancy. Above 100 mGy abortion might be considered, especially if the unborn child received the dose during the first trimester of gestation. For proper management of pregnant patients exposed to X-rays, medical physicists need to estimate the conceptus dose. Chapter 5 describes methods for conceptus dose estimation from diagnostic and interventional X-ray examinations.

The background risk of childhood fatal cancer is about 0.2%. If the conceptus dose from an X-ray examination is 10 mGy the risk of excess childhood fatal cancer is 0.06%. This risk estimation is based on the risk coefficient of 6% Sv^{-1}. This is a very small increase in the probability of fatal cancer. If the conceptus dose is 100 mGy the probability of cancer goes to 0.8%. This is also a low probability of radiation-induced cancer and, for this reason, abortion should not be considered if the dose to the unborn child is lower than 100 mGy. However, when the conceptus dose is above 100 mGy deterministic effects are likely to be observed and, for this reason, abortion might be considered especially if the exposure took place during the first trimester.

Communication of medical physicists with radiologists and referring physicians
Medical physicists should estimate conceptus doses and risks and provide a report to the radiologists and referring physicians. The main sections of this report are presented below.

1. A table presenting the most important exposure parameters;
2. A short description of the method(s) used to estimate conceptus dose;
3. Conceptus dose results;
4. Information about conceptus radiogenic risk;
5. Radiation dose optimization strategies.

Communication with the pregnant patient

Communication with the patient is the key component of a program developed to manage a pregnant patient who requires radiological examination. Pregnant patients should be properly informed about expected benefits and potential risks so that they can make informed choices. Healthcare providers should have sufficient knowledge to communicate conceptus risks sufficiently. A literature review shows that knowledge of radiation dose and risk is poor among referring physicians, radiologists, radiographers, trainees and medical students (Ramanathan and Ryan 2015, Paolicchi *et al* 2016, Szarmach *et al* 2015, Badawy *et al* 2015). A study found that radiologists tend to underestimate patient radiation doses from common X-ray examinations (Ramanathan and Ryan 2015). This is of serious concern, as it may lead to (a) provision of inaccurate information to patients about the risks associated with radiation doses and (b) acceptance of many unjustified X-ray procedures. Another research work shows that radiographers also need to improve their knowledge in medical radiation protection (Paolicchi *et al* 2016). Most of the radiographers surveyed underestimated the doses of almost all X-ray examinations. It is encouraging, however, that young radiographers showed a better knowledge compared with the more experienced radiographers. A questionnaire was distributed to physicians, nurses and other personnel who use ionizing radiation (Szarmach *et al* 2015). This study shows that systematic training courses for medical personnel must be considered to increase awareness in medical radiation protection. A study was carried out in Australia to assess the radiation awareness amongst physicians in emergency departments (Badawy *et al* 2015). Most of the physicians surveyed reported that they would not be confident discussing radiogenic risks with patients and indicated the need for additional education.

Ionizing radiation is frightening because, for patients and relatives, radiation is often linked to atomic bombs and nuclear reactor accidents. Proper communication about the risks from medical X-ray examinations between healthcare providers and pregnant patients is needed to increase understanding and eliminate unfounded fears. Dauer *et al* (2011) discuss four approaches typically used by medical professionals to communicate risks of radiation to patients: the 'paternalistic' approach ('the physician advises the patient what procedures and treatments are recommended and the patient is expected to unquestionably follow such advice'), the 'risk comparisons only' approach ('risk comparisons using the concept of effective dose'), the 'risk numerology' approach ('relative risks, excess cancer rates, increased rates over background levels, and log-based hazard comparisons') and the 'quality assurances' approach ('our protocols are designed to deliver doses as low as reasonably achievable so you don't really need to worry about it'). In the same publication, authors provide suggestions for improved benefit-and-risk communication.

When communicating with a pregnant patient who needs a diagnostic or interventional X-ray examination, it is sometimes helpful to compare conceptus dose with that from background radiation over a period of years or with exposure during flights. To take the right decision, the pregnant patient should also know the risk of not having the X-ray examination. Information about conditions present at birth regardless of radiation shall also be given. There is always a chance that a woman can give birth to a congenitally malformed child regardless of any exposure to radiation. The current incidence of serious birth defects in human populations is about 6%. After the benefit–risk dialogue, the pregnant patient should be asked to sign a consent form.

Protocol optimization

Once a decision has been made that a diagnostic or interventional procedure should be performed on a pregnant patient using X-rays, the exposure has to be optimized. Chapter 6 provides guidelines and practical hints for the optimization of radiologic procedures performed on pregnant patients.

8.1.3 Accidental exposure of pregnant patients

Pregnancy screening

All patients of childbearing age, that is from 12 to 50 years old, should be screened about pregnancy status prior to X-ray imaging preferably using a standardized form or verbally. Referring physicians must check a box on the examination request form if the patient is pregnant. X-ray departments must have posters in the waiting area and other suitable areas asking patients to inform the radiographer or the radiologist before commencement of the examination about a possible pregnancy. Menstrual history is not always reliable in determination of pregnancy, especially in patients with irregular periods (Damilakis 2004). Investigation of pregnancy status should be thorough and, if needed, should include questions related to contraceptive and sexual history.

Each facility should develop a policy to address pregnancy screening for patients undergoing radiologic examinations taking into account scientific evidence and national legislation. This policy should address a number of issues including (a) how pregnancy screening should be performed, (b) when and how a pregnancy test should be done, (c) how to protect women's privacy during screening, (d) how to screen adolescent girls and (e) how to screen patients when an emergent X-ray examination is needed. Pregnancy tests cannot completely exclude the possibility of pregnancy during the first post-conception weeks. ICRP (2000) states that it is prudent to consider as pregnant any woman of reproductive age presenting herself for an X-ray examination at a time when a menstrual period is overdue, or missed, unless there is information that precludes a pregnancy.

The 10-day rule was developed in 1959 to limit abdominal examinations to the first 10 days from the onset of the menstrual cycle. This rule may not provide sufficient protection to those with very short menstrual cycles. Moreover, it is known today that diagnostic or interventional X-rays cannot cause damage to the conceptus during the first post-conception weeks. In a few extreme cases, high doses from these procedures might terminate pregnancy ('all-or-nothing' effect). This rule can cause organization and scheduling problems and should not be applied routinely. According to the 28-day rule, an X-ray examination may be performed if the female patient has had a period within the last 28 days. This rule is based on the 28-day menstrual cycle. However, the length of the cycle varies considerably and, for this reason, the 28-day rule has been abandoned.

Management of pregnant patient exposed accidentally

In the case of accidental irradiation, risk assessment and proper risk communication is of great importance. Medical physicists should estimate conceptus dose and assess radiogenic risk taking into consideration conception age. To estimate conceptus dose, medical physicists must gather certain information to recreate the event. No assessment of conceptus dose can be reasonably made without knowing the equipment where the exposure took place and the exact protocol parameters used during the study.

After the examination, the pregnant patient whose embryo was exposed accidentally is usually very anxious and, sometimes, she considers the possibility of abortion. These female patients must be given complete and honest information. A study (Ratnapalan *et al* 2004) has found that 'physicians who care for pregnant women perceive the teratogenic risk associated with an abdominal radiograph and an abdominal CT scan to be unrealistically high during early pregnancy. This misperception could lead to increased anxiety among pregnant women seeking counseling and to unnecessary terminations of otherwise wanted pregnancies. This perception of high teratogenic risk associated with radiation could also lead to delay in needed care of pregnant women'. Training courses are needed to provide healthcare staff with the necessary knowledge for proper management of pregnant patients needing X-ray diagnostic and interventional procedures.

8.2 The management of pregnant employees

A topic of great interest for medical physicists is the dose management of pregnant employees working in diagnostic and interventional radiology. The following sections describe the dose limit proposed by the European Union and the ICRP regarding the protection of the unborn child of an occupationally exposed pregnant woman and a program to control conceptus exposure to ionizing radiation.

8.2.1 Program to evaluate and control conceptus dose

Special recommendations and guidelines apply after a pregnant employee declares her pregnancy. The ICRP (ICRP 2000) states that *'The working conditions of a pregnant worker, after the declaration of pregnancy, should be as such to make it unlikely that the additional dose to the conceptus will exceed about 1 mGy during the remainder of pregnancy'.* The new European BSS (2013) states that

> *'1. Member States shall ensure that the protection of the unborn child is comparable with that provided for members of the public. As soon as a pregnant worker informs the undertaking or, in the case of an outside worker, the employer, of the pregnancy, in accordance with national legislation the undertaking, and the employer, shall ensure that the employment conditions for the pregnant worker are such that the equivalent dose to the unborn child is as low as reasonably achievable and unlikely to exceed 1 mSv during at least the remainder of the pregnancy.*

2. As soon as workers inform the undertaking, or in case of outside workers, the employer, that they are breastfeeding an infant, they shall not be employed in work involving a significant risk of intake of radionuclides or of bodily contamination'.

The time period from conception until declaration of pregnancy is covered by the normal protection of workers (ICRP 2000). The '1 mSv equivalent dose limit to the unborn child' is equivalent to a dose limit of approximately 2 mSv to the surface of the woman's abdomen for the remainder of the pregnancy, once it has been declared. Radiation doses to occupationally exposed healthcare personnel from many fluoroscopy examinations and fluoroscopically-guided interventional procedures are high, and it is likely that in some cases, for a pregnant employee, the conceptus equivalent dose limit of 1 mSv can be approached or even exceeded. Early declaration of pregnancy is needed in view of the conceptus risks associated with radiation exposure.

The most important elements of a program suitable for the dose management of pregnant personnel are (a) declaration of pregnancy, (b) evaluation of the working conditions of the pregnant worker, (c) conceptus dose anticipation and workload estimation, (d) counseling and (e) dose monitoring. From the radiation protection point of view, a radiation worker is considered pregnant only after declaration of pregnancy. A pregnancy should be formally declared in writing. Following the submittal of a declaration of pregnancy, the working conditions of the pregnant worker should be evaluated. Medical physicists together with the pregnant woman must evaluate the work situation taking into account doses received during the previous years. Personal dose monitoring will show whether the woman is receiving measurable radiation doses. This is important information for medical physicists to decide whether conceptus dose anticipation and workload estimation is needed.

8.2.2 Conceptus dose anticipation and workload estimation

For the majority of pregnant workers, conceptus dose determination is not needed. For example, conceptus dose determination is not normally needed for a technologist responsible for radiographic exposures with a fixed radiographic unit. However, staff carrying out fluoroscopic and interventional procedures may receive a considerable amount of radiation. For low dose procedures, conceptus dose anticipation is a very important element of the dose management program of the pregnant worker.

Conceptus dose anticipation and maximum workload determination is needed for pregnant staff participating in fluoroscopy and fluoroscopically-guided interventional procedures. Personal dosimeters are of limited value because they can provide only the dose to the surface of the woman's abdomen, retrospectively. Iso-dose maps of air kerma doses normalized to the dose-area product (DAP) for projections and procedures commonly employed in the electrophysiological laboratory have been published (Damilakis 2005). This information can be used to (a) anticipate conceptus dose from occupational exposure of pregnant staff during these fluoroscopically-guided procedures and (b) estimate the maximum workload allowed for each month of gestation period following pregnancy declaration. Information provided in the above publication is also useful for young female employees who are planning to become pregnant. These workers could take into consideration iso-dose maps and select the most suitable position that allows them to decrease radiation doses to the pelvic area as much as possible (Damilakis 2005).

The CoDE web-based software tool (http://embryodose.med.uoc.gr/code/) provides estimates of conceptus absorbed dose of a pregnant employee who participates in fluoroscopically-guided interventional procedures and its use is free of charge. A C-arm unit and an anthropomorphic phantom simulating an average patient were used to determine scatter exposure rates at specific locations over a 50 cm × 50 cm grid around the table of the angiography suite. Data were collected for all commonly used fluoroscopic projections centered on three different anatomical regions, that is chest, abdomen or pelvis. These projections are shown in table 5.5, chapter 5. For each projection, exposure data were obtained for various combinations of tube voltage and total filtration. Exposures in μSv/h were measured at the operator's waist that is 110 cm from the floor. Projection-specific spatial 2-d maps of normalized to DAP scatter exposure rate were derived.

Scatter exposure data were obtained for various fluoroscopy beam field sizes. Monte Carlo simulation was employed to determine the exposure reduction factor achieved by using a radioprotective apron for different values of operating tube voltage and protective apron lead equivalence. Scatter exposures at the specific position of the pregnant employee in the operating room was converted to conceptus dose using gestation stage-specific air kerma to embryo dose (*ED*) conversion factors previously published (Damilakis *et al* 2005).

ED from a specific procedure for which *n* different projections are involved was calculated from:

$$ED = \sum_i^n (NE_i(\mathrm{p,\ kV,\ filtration}) \times \mathrm{DAP}_i \times f_{\mathrm{field\ size}} \times f_{\mathrm{gest.\ stage}} \times f_{\mathrm{Pb\ apron}})$$

where NE_i is the normalized scatter exposure at the waist level for the fluoroscopic projection *i* determined for the same tube voltage and total beam filtration, DAP_i is the cumulative DAP recorded for the specific projection *i*, $f_{\mathrm{field\ size}}$ is the correction factor for the specific beam field size at entrance skin surface, $f_{\mathrm{gest.\ stage}}$ is the correction factor for the selected gestational stage of the pregnant employee, and $f_{\mathrm{Pb\ apron}}$ is the correction factor for the specific lead apron worn by the pregnant employee.

8.2.3 Counseling and dose monitoring

Pregnant workers should be given information and instructions to cover the fundamental and routine requirements. They should be informed of the dose limits, potential doses from everyday clinical practice and potential radiation-induced risks. An individualized radiation safety plan should be developed for the expectant mother taking into consideration her specific working conditions. Work or area restrictions include prohibiting access to certain areas or restricting performance of certain skills. However, reduction of the number of procedures is the most important work restriction for the pregnant worker working in an interventional suite.

Pregnancy discrimination must be avoided. The expectant mother can choose to perform her job without any change in assigned working duties. Sometimes, employers provide pregnant workers the possibility to change their position to another where the radiation exposure is lower. The option of a position that does not expose pregnant workers to occupational ionizing radiation is also sometimes offered. The final decision should be made taking into consideration several parameters including expectant mother's request, local policies and national regulations.

Radiation protection equipment including maternity lead aprons should be available. Pregnant workers are expected to use them regularly and perform their duties adhering to the as-low-as-reasonably-achievable (ALARA) principle. Regarding dose monitoring, a separate badge for fetal monitoring should be worn beneath the shielding at waist level in addition to the standard whole-body personal dosimeter. Direct-reading digital alarming dosimeters should also be available. These dosimeters are useful, especially during the first month after declaration of pregnancy. During this time period, the expectant mother might have doubts about the radiation protection of her child, so these dosimeters may reduce her anxiety.

References

Badawy M K, Sayakkarage D and Ozmen M 2015 Awareness of radiation dose associated with common diagnostic procedures in emergency departments: a pilot study *AMJ* **8** 338–44

Damilakis J 2004 Pregnancy and diagnostic X-rays *Radiation Protection* ed J Damilakis (Berlin Heidelberg: Springer) pp 33–9

Damilakis J, Perisinakis K, Theocharopoulos N, Tzedakis A, Manios E, Vardas P and Gourtsoyiannis N 2005 Anticipation of radiation dose to the embryo from occupational exposure of pregnant staff during fluoroscopically guided electrophysiological procedures *J. Cardiovasc. Electrophysiol.* **16** 773–80

Dauer L, Thornton R, Hay J, Balter R, Williamson M and Germain J 2011 Fear, feelings and facts: interactively communicating benefits and risks of medical radiation with patients *AJR* **196** 756–61

European BSS 2013 Council Directive 2013/59/Euratom laying down basic safety standards for protection against the dangers arising from exposure to ionising radiation, and repealing Directives 89/618/Euratom, 90/641/Euratom, 96/29/Euratom, 97/43/Euratom and 2003/122/Euratom. Official Journal L-13 of 17.01.2014

Goldman S M and Wagner L K 1996 Radiologic management of abdominal trauma in pregnancy *AJR* **166** 763–7

ICRP 2000 Publication 84 *Pregnancy and Medical Radiation* (Exeter: Elsevier Science Ltd)

Paolicchi F, Miniati F and Bastiani L *et al* 2016 Assessment of radiation protection awareness and knowledge about radiological examination doses among Italian radiographers *Insights Imaging* **7** 233–42

Ramanathan S and Ryan J 2015 Radiation awareness among radiology residents, technologists, fellows and staff: where do we stand? *Insights Imaging* **6** 133–9

Ratnapalan S, Bona N, Chandra K and Koren G 2004 Physicians' perceptions of teratogenic risk associated with radiography and CT during early pregnancy *AJR* **182** 1107–9

Szarmach A, Piskunowicz M and Swieton D *et al* 2015 Radiation safety awareness among medical staff *Pol. J. Radiol.* **80** 57–61

Wagner L K, Lester R G and Saldana L R 1997 Exposure of the pregnant patient to diagnostic radiations *A Guide to Medical Management* 2nd edn (Madison, WI: Medical Physics Publishing)

IOP Publishing

Appendix A

Dose quantities for measurements

Radiography or fluoroscopy:

1. Incident air kerma (Ki) in Gy.
   This is the kerma to air from an incident X-ray beam measured on the central beam axis at the position of the patient or phantom surface. Only the radiation incident on the patient or phantom and not the backscattered radiation is included.
2. Entrance surface air kerma (ESAK) in Gy.
   The ESAK is the kerma to air measured on the central beam axis at the position of the patient or phantom surface. The radiation incident on the patient or phantom and the backscattered radiation are included.
3. X-ray tube output (X) in Gy mAs^{-1}.
   The X-ray tube output is the quotient of the air kerma at a specified distance, *d*, from the X-ray tube focus by the tube current–exposure time product (mAs).
4. Air kerma–area product (KAP) known also as dose area product (DAP) in $Gy{\cdot}m^2$.
   The air kerma–area product is the integral of the air kerma over the area of the X-ray beam in a plane perpendicular to the beam axis.

CT dosimetry: There are specific dose descriptors in CT and these are: (1) computerized tomography dose index (CTDI), (2) dose length product (DLP), and (3) size-specific dose estimate (SSDE).

Computerized tomography dose index:

- CTDI integrates the radiation dose imparted within and beyond a single slice and it is defined by the following equation:

doi:10.1088/978-0-7503-1317-9ch9

$$\mathrm{CTDI} = \frac{1}{T}\int_{-\infty}^{+\infty} D(z)dz \tag{A.1}$$

T is the nominal slice thickness, $D(z)$ is the dose profile along a line parallel to the z-axis (tube rotation axis).

- CTDI_w is used for approximating the average dose over a single slice in order to account for variations in dose values between the center and the periphery of the slice. It is defined by the following equation:

$$\mathrm{CTDI}_w = \frac{1}{3}\mathrm{CTDI}_c + \frac{2}{3}\mathrm{CTDI}_p \tag{A.2}$$

CTDI_p is the average of the four CTDI_p values measured in the periphery of the phantom (12, 3, 6 and 9 o'clock).

- $\mathrm{CTDI}_{\mathrm{vol}}$ represents the radiation dose in one tube rotation in MDCT and allows for variations in exposure in the z direction when the pitch (p) is not equal to 1 (pitch is the ratio of table feed in one rotation (I) to slice collimation (NT)).

$$\mathrm{CTDI}_{\mathrm{vol}} = \frac{\mathrm{NT}}{I} \times \mathrm{CTDI}_w \tag{A.3}$$

$$\mathrm{CTDI}_{\mathrm{vol}} = \frac{\mathrm{CTDI}_w}{p} \tag{A.4}$$

This equation applies when p is not equal to 1.

CTDI is measured in mGy and display of the CTDI value on the CT console is strongly recommended. For CTDI measurement, two polymethylmethacrylate (PMMA) cylinders of 14 cm length can be used. For head examinations, a phantom diameter of 16 cm is used and for the body a phantom diameter of 32 cm is applied. The phantoms are called, respectively, the head and body CTDI phantoms. CTDI is measured using a specially designed pencil ionization chamber with an active length of 100 mm both in free air at the center of rotation ($\mathrm{CTDI}_{\mathrm{air}}$) and within the holes of the two phantoms. CTDI_c and CTDI_p are defined, respectively, as the CTDI values measured with a pencil chamber dosemeter positioned in the center and in the periphery of the PMMA head or body phantom.

It should be noted that CTDI has a few limitations. It is measured by using a standardized, homogeneous, cylindrical phantom and therefore it possibly differs from the dose for objects of substantially different size, shape, or attenuation, like the human body. It is expressed as dose to air, not dose to tissue and it is not sufficient for slice collimations greater than 10 cm such as those of 256 or 320 CT scanners. Finally, it does not indicate the dose to a specific point in the scan volume when the patient table remains stationary for multiple scans, such as for interventional or perfusion CT.

Dose length product: DLP is used to calculate the dose for a series of slices or a complete examination and is defined by the following equation:

$$\mathrm{DLP} = \sum_{i}^{N} \mathrm{CTDI}_w \mathrm{TN}$$

i represents each one of the individual N scans of the examination that covers a length T of patient anatomy. It is a way to evaluate the total radiation dose given to the patient during a specific examination. This practically means that for a given technical protocol with certain $\mathrm{CTDI}_{\mathrm{vol}}$, the DLP of two scanning regions with different lengths will be different.

Size-specific dose estimate: Size-specific dose estimate (SSDE) is a newer CT-derived value that incorporates patient size as a modifying correction factor to better estimate patient dose. The methodology to estimate SSDE from $\mathrm{CTDI}_{\mathrm{vol}}$ is described in detail within the AAPM reports 204 and 220 (AAPM 2011, 2014). SSDE is determined by multiplying $\mathrm{CTDI}_{\mathrm{vol}}$ by a conversion factor based on the patient's effective diameter. The effective diameter is defined as the square root of the product of the anterior-posterior and lateral patient diameter. Limited scanners display SSDE on the CT console. It can, however, be calculated using patient localizer images or other methodology as described in AAPM report 220.

Dose quantities related to stochastic and deterministic effects
Organ or tissue dose: The mean absorbed dose in a specified tissue or organ is given the symbol D_T and is equal to the ratio of the energy imparted to the tissue or organ (e) to the mass of the tissue or organ.
Equivalent dose in Sv: For a single type of radiation (R), equivalent dose (H_T) is the product of a radiation weighting factor (w_R), for radiation R and organ dose (D_T), thus:

$$H_T = w_R \, D_T$$

The unit of equivalent dose is sievert (Sv). The radiation weighting factor, w_R, allows for differences in the relative biological effectiveness of the incident radiation in producing stochastic effects at low doses in tissue or organ, T. For X-ray energies used in diagnostic radiology, w_R is taken to be unity (A.1).

Effective dose (E) in Sv: The effective dose (E) is a concept used to normalize partial body irradiations relative to whole body irradiations to enable comparisons of risk. Its estimation requires knowledge of the dose to specific sensitive organs within the body, which are typically obtained from Monte Carlo modeling of absorbed organ doses within mathematical anthropomorphic phantoms and recently also voxel phantoms derived from CT scans of human cadavers. It is expressed in milliSieverts (mSv) and can be compared to the effective dose from other sources of ionizing radiation, such as that from background radiation level (e.g., radon, cosmic radiation, etc) or nuclear medicine diagnostic examinations.

References

AAPM 2011 Report 204 Size-specific dose estimates (SSDE) in paediatric and adult body CT examinations, AAPM

AAPM 2014 Report 220 Use of water equivalent diameter for calculating patient size and size-specific dose estimates (SSDE) in CT, AAPM

www.ingramcontent.com/pod-product-compliance
Ingram Content Group UK Ltd.
Pitfield, Milton Keynes, MK11 3LW, UK
UKHW051856100926
8556IPUK00005B/648

9 780750 313186